FE/EIT EXAM PR

CHEMICAL ENGINEERING

Third Edition

Dilip K. Das, MS, PE Chem. Eng. & Rajaram K. Prabhudesai, PhD, PE Chem. Eng.

This publication is designed to provide accurate and authoritative information in regard to the subject matter covered. It is sold with the understanding that the publisher is not engaged in rendering legal, accounting, or other professional service. If legal advice or other expert assistance is required, the services of a competent professional person should be sought.

President: Roy Lipner
Vice President & General Manager: David Dufresne
Vice President of Product Development and Publishing: Evan M. Butterfield
Editorial Project Manager: Laurie McGuire
Director of Production: Daniel Frey
Production Editor: Caitlin Ostrow
Creative Director: Lucy Jenkins

Copyright 2008 by Dearborn Financial Publishing, Inc.®

Published by Kaplan AEC Education
30 South Wacker Drive
Chicago, IL 60606-7481
(312) 836-4400
www.kaplanaecengineering.com

All rights reserved. The text of this publication, or any part thereof, may not be reproduced in any manner whatsoever without written permission in writing from the publisher.

Printed in the United States of America.

08   10   9   8   7   6   5   4   3   2   1

CONTENTS

Introduction ix

About the Authors xvii

CHAPTER 1
Dimensions and Units 1
UNITS OF FORCE 1
MOLAR UNITS 2
MULTIPLIERS AND CONVERSION FACTORS 2
REFERENCES 4

CHAPTER 2
Material and Energy Balances 11
ACCOUNTING PRINCIPLE 12
DEFINITIONS AND RELATIONSHIPS OF MATERIAL PROPERTIES 12
VAPOR PRESSURE 13
IDEAL GASES 14
IDEAL GAS MIXTURES 15
CRITICAL PROPERTIES 15
NON-IDEAL OR REAL GASES 15
SOLUTIONS 16
HUMIDITY AND SATURATION 17
SOLUBILITY AND CRYSTALLIZATION 18
MASS BALANCES 18
ENERGY BALANCES 19
LAW OF THERMOCHEMISTRY 21
FUELS AND COMBUSTION 21
REFERENCES 22

CHAPTER 3
Thermodynamics 29
TERMINOLOGY 30
THERMODYNAMIC PROPERTIES 30
SECOND LAW OF THERMODYNAMICS 32
PROPERTIES OF IDEAL GASES 35
PROPERTIES OF REAL GASES 36
THIRD LAW OF THERMODYNAMICS 42
CYCLIC PROCESSES 42
HEATS OF MIXING AND ENTHALPY OF A SOLUTION 45
GIBBS' PHASE RULE 46
REFERENCES 46

CHAPTER 4

Mass Transfer 55

FICK'S LAW 55

MASS TRANSFER UNDER NON-FLOW CONDITIONS 57

MASS TRANSFER COEFFICIENTS 58

MASS TRANSFER FROM A GAS INTO A FALLING LIQUID FILM 59

MASS TRANSFER FROM SPHERES 59

TURBULENT MASS TRANSFER 60

INTERPHASE MASS TRANSFER 61

OVERALL MASS TRANSFER COEFFICIENTS 62

MASS TRANSFER IN PACKED BEDS 63

DIFFUSION IN SOLIDS 65

REFERENCES 66

CHAPTER 5

Chemical Kinetics 71

THERMODYNAMICS OF CHEMICAL REACTIONS 72

CLASSIFICATION OF CHEMICAL REACTIONS 73

CHEMICAL EQUILIBRIUM 73

RATE OF CHEMICAL REACTION (HOMOGENEOUS REACTIONS) 75

MOLECULARITY AND ORDER OF REACTION 76

IRREVERSIBLE REACTIONS 76

REVERSIBLE HOMOGENEOUS REACTIONS 76

COMPLEX REACTIONS 78

PARALLEL REACTIONS 78

CONSECUTIVE REACTIONS 79

HOMOGENEOUS CATALYZED REACTIONS 79

AUTOCATALYTIC REACTIONS 80

RATE EXPRESSIONS IN TERMS OF FRACTIONAL CONVERSION X 80

MICHAELIS-MENTON EQUATION 82

CONSTANTS OF THE RATE EQUATIONS 82

METHOD OF k CALCULATION 83

EFFECT OF TEMPERATURE ON RATE OF REACTION 84

REACTOR DESIGN FOR HOMOGENEOUS REACTIONS 84

BATCH REACTORS 84

REFERENCES 87

CHAPTER 6

Process Design and Economic Evaluation 95

DEGREES OF FREEDOM FOR OPTIMIZATION 95

FIXED INVESTMENT 96

GROSS PROFIT AND NET PROFIT 96

RETURN ON INVESTMENT 96

PAYOUT OR PAYBACK TIME 97

VENTURE PROFIT AND VENTURE COST 97

LINEAR BREAK-EVEN ANALYSIS 97

ORDER-OF-MAGNITUDE ESTIMATE OF EQUIPMENT AND INSTALLED COSTS 97

CHAPTER 7 Heat Transfer 103

CONDUCTION 103

CONVECTION 105

RADIATION 105

FLUID-TO-FLUID HEAT TRANSFER ACROSS A SOLID WALL 107

UNSTEADY STATE HEAT TRANSFER 111

REFERENCES 114

CHAPTER 8 Transport Phenomena 121

DENSITY, SPECIFIC VOLUME, SPECIFIC WEIGHT, AND SPECIFIC GRAVITY 122

VISCOSITY 122

STATIC PRESSURE, STATIC HEAD 123

STEADY, INCOMPRESSIBLE FLOW OF FLUID IN CONDUITS AND PIPES 124

EQUIVALENT DIAMETER FOR NONCIRCULAR CONDUITS 126

FRICTION FACTOR, NEWTONIAN FLUID 127

EQUIVALENT LENGTH OF A PIPING SYSTEM 128

RESISTANCE COEFFICIENT, K, OF A FITTING 128

DRAG COEFFICIENT 130

COMPRESSIBLE FLOW 130

TEMPERATURE RISE DUE TO SKIN FRICTION UNDER ADIABATIC CONDITION 132

REFERENCES 132

CHAPTER 9 Process Control 139

CONTROL SYSTEMS 139

BLOCK DIAGRAMS 142

TRANSFER FUNCTIONS 142

PROPORTIONAL BAND 143

TRANSIENT RESPONSE OF CONTROL SYSTEMS 144

FIRST AND SECOND ORDER SYSTEMS 145

REFERENCES 146

CHAPTER 10 Process Equipment Design 151

FLUID HANDLING EQUIPMENT 151

HEAT TRANSFER EQUIPMENT 156

PRESSURE VESSELS 156

vi   Contents

MASS TRANSFER COLUMNS 157
REFERENCES 158

CHAPTER 11 Computer and Numerical Methods 167
INTRODUCTION 168
SPREADSHEETS 173
NUMERICAL METHODS 174
NUMERICAL INTEGRATION 176
NUMERICAL SOLUTIONS OF DIFFERENTIAL EQUATIONS 177
PACKAGED PROGRAMS 178
GLOSSARY OF COMPUTER TERMS 178

CHAPTER 12 Process Safety 187
THRESHOLD LIMIT VALUES 187
FIRE AND EXPLOSION ISSUES 189
CLASSIFICATION OF LIQUIDS FOR THERMAL HAZARDS ANALYSIS 191
COMPUTATION OF FLASH POINT 192
CALCULATING FLAMMABILITY LIMITS 192
STOICHIOMETRY OF COMBUSTION REACTION AND ESTIMATION OF FLAMMABILITY LIMITS 193
INERTING AND PURGING 194
LIMITING AND EXCESS REACTANT 195
REFERENCES 195

CHAPTER 13 Pollution Prevention 199
FEDERAL POLLUTION PREVENTION ACT OF 1990 200
TERMINOLOGY 200
OZONE, FRIEND AND FOE 203
ABATEMENT OF AIR POLLUTION 204
ABATEMENT OF WATER POLLUTION 206
TREATMENT OF CONTAMINATED SOIL 207
REFERENCES 207

CHAPTER 14 Distillation 211
VAPOR-LIQUID EQUILIBRIA 212
IDEAL SYSTEMS 215
RELATIVE VOLATILITY 216
NONIDEAL SYSTEMS 217
EQUILIBRIUM VAPORIZATION RATIOS (EQUILIBRIUM CONSTANTS) K 218
BUBBLE POINT 218

FLASH (EQUILIBRIUM) DISTILLATION 219
DIFFERENTIAL BATCH DISTILLATION 220
FRACTIONAL DISTILLATION 220
DESIGN OF COLUMNS FOR SIMPLE BINARY SYSTEMS 222
REFERENCES 229

APPENDIX **Afternoon Sample Examination 237**

Index 267

Introduction

OUTLINE

HOW TO USE THIS BOOK   IX

BECOMING A PROFESSIONAL ENGINEER   X
Education ■ Fundamentals of Engineering/Engineer-In-Training Examination ■ Experience ■ Professional Engineer Examination

FUNDAMENTALS OF ENGINEERING/
ENGINEER-IN-TRAINING EXAMINATION   XI
Examination Development ■ Examination Structure ■ Taking the Examination ■ Examination Procedure ■ Examination-Taking Suggestions ■ License Review Books ■ Textbooks ■ Examination Day Preparations ■ Items to Take to the Examination ■ Special Medical Condition ■ Examination Scoring

HOW TO USE THIS BOOK

Chemical Engineering: FE/EIT Exam Preparation is designed to help you prepare for the Fundamentals of Engineering/Engineer-in-Training exam. The book covers the discipline-specific afternoon exam in chemical engineering. For the morning exam, Kaplan AEC offers the comprehensive review book, *Fundamentals of Engineering: FE/EIT Exam Preparation*.

This book covers the major topics on the afternoon exam in chemical engineering, reviewing important terms, equations, concepts, analysis methods, and typical problems. After reviewing the topic, you can work the end-of-chapter problems to test your understanding. Complete solutions are provided so that you can check your work and further refine your solution methodology.

After reviewing individual topics, you should take the Sample Exam at the end of the book. To fully simulate the exam experience, you should answer these 60 questions in an uninterrupted four-hour period, without looking back at any content in the rest of the book. You may wish to consult the *Fundamentals of Engineering Supplied-Reference Handbook*, which is the only reference you are allowed to use in the actual exam.

When you've completed the Sample Exam, check the provided solutions to determine your correct and incorrect answers. This should give you a good sense of topics you may want to spend more time reviewing. Complete solution methods are shown, so you can see how to adjust your approach to problems as needed.

The following sections provide you with additional details on the process of becoming a licensed professional engineer and on what to expect at the exam.

BECOMING A PROFESSIONAL ENGINEER

To achieve registration as a Professional Engineer, there are four distinct steps: (1) education, (2) the Fundamentals of Engineering/Engineer-in-Training (FE/EIT) exam, (3) professional experience, and (4) the professional engineer (PE) exam. These steps are described in the following sections.

Education

Generally, no college degree is required to be eligible to take the FE/EIT exam. The exact rules vary, but all states allow engineering students to take the FE/EIT exam before they graduate, usually in their senior year. Some states, in fact, have no education requirement at all. One merely need apply and pay the application fee. Perhaps the best time to take the exam is immediately following completion of related coursework. For most engineering students, this will be the end of the senior year.

Fundamentals of Engineering/ Engineer-In-Training Examination

This eight-hour, multiple-choice examination is known by a variety of names—Fundamentals of Engineering, Engineer-in-Training (EIT), and Intern Engineer—but no matter what it is called, the exam is the same in all states. It is prepared and graded by the National Council of Examiners for Engineering and Surveying (NCEES).

Experience

States that allow engineering-seniors to take the FE/EIT exam have no experience requirement. These same states, however, generally will allow other applicants to substitute acceptable experience for coursework. Still other states may allow a candi-date to take the FE/EIT exam without any education or experience requirements.

Typically, four years of acceptable experience is required before one can take the Professional Engineer exam, but the requirement may vary from state to state.

Professional Engineer Examination

The second national exam is called Principles and Practice of Engineering by NCEES, but many refer to it as the Professional Engineer exam or PE exam. All states, plus Guam, the District of Columbia, and Puerto Rico, use the same NCEES exam. Review materials for this exam are found in other engineering license review books.

FUNDAMENTALS OF ENGINEERING/ENGINEER-IN-TRAINING EXAMINATION

Laws have been passed that regulate the practice of engineering in order to protect the public from incompetent practitioners. Beginning in 1907 the individual states began passing *title* acts regulating who could call themselves engineers and offer services to the public. As the laws were strengthened, the practice of engineering was limited to those who were registered engineers, or to those working under the supervision of a registered engineer. Originally the laws were limited to civil engineering, but over time they have evolved so that the titles, and some-times the practice, of most branches of engineering are included.

There is no national licensure law; licensure is based on individual state laws and is administered by boards of registration in each state. You can find a list of contact information for and links to the various state boards of registration at the Kaplan AEC Web site: *www.kaplanaecengineering.com*. This list also shows the exam registration deadline for each state.

Examination Development

Initially, the states wrote their own examinations, but beginning in 1966 NCEES took over the task for some of the states. Now the NCEES exams are used by all states. Thus it is easy for engineers who move from one state to another to achieve licensure in the new state. About 50,000 engineers take the FE/EIT exam annually. This represents about 65% of the engineers graduated in the United States each year.

The development of the FE/EIT exam is the responsibility of the NCEES Committee on Examination for Professional Engineers. The committee is composed of people from industry, consulting, and education, all of whom are subject-matter experts. The test is intended to evaluate an individual's understanding of mathematics, basic sciences, and engineering sciences obtained in an accredited bachelor degree of engineering. Every five years or so, NCEES conducts an engineering task analysis survey. People in education are surveyed periodically to ensure the FE/EIT exam specifications reflect what is being taught.

The exam questions are prepared by the NCEES committee members, subject matter experts, and other volunteers. All people participating must hold professional licensure. When the questions have been written, they are circulated for review in workshop meetings and by mail. You will see mostly metric units (SI) on the exam. Some problems are posed in U.S. customary units (USCS) because the topics typically are taught that way. All problems are four-way multiple choice.

Examination Structure

The FE/EIT exam is divided into a morning four-hour section and an afternoon four-hour section. There are 120 questions in the morning section and 60 in the afternoon.

The morning exam covers the topics that make up roughly the first $2\frac{1}{2}$ years of a typical engineering undergraduate program.

Seven different exams are in the afternoon test booklet, one for each of the following six branches: civil, mechanical, electrical, chemical, industrial, environmental. A general exam is included for those examinees not covered by the six engineering branches. Each of the six branch exams consists of 60 problems covering coursework in the specific branch of engineering. The general exam, also 60 problems, has topics that are similar to the morning topics.

If you are taking the FE/EIT as a graduation requirement, your school may compel you to take the afternoon exam that matches the engineering discipline in which you are obtaining your degree. Otherwise, you can choose the afternoon exam you wish to take. There are two approaches to deciding.

One approach is to take the general afternoon exam regardless of your engineering discipline. Because the topics covered are similar to those in the morning exam, this approach may streamline your review time and effort. If you are still in college or recently graduated, these general topics may be very fresh in your mind.

The second approach is to take the afternoon exam that matches the discipline you majored in. Particularly if you have been out of college for several years, practicing this discipline in your daily work, you will be very familiar and comfortable with the topics. This may be to your advantage during your review time and in the pressure of the exam itself.

At the beginning of the afternoon test period, examinees will mark the answer sheet as to which branch exam they are taking. You could quickly scan the test, judge the degree of difficulty of the general versus the branch exam, then choose the test to answer. We do not recommend this practice, as you would waste time in determining which test to write. Further, you could lose confidence during this indecisive period.

Table 1 summarizes the major subjects for the chemical engineering afternoon exam, including the percentage of problems you can expect to see on each one.

Taking the Examination

The National Council of Examiners for Engineering and Surveying (NCEES) prepares FE/EIT exams for use on a Saturday in April and October each year.

Table 1 Chemical Engineering Afternoon Exam

NCEES Defined Topics	Percentage of Problems	Review Chapter in Text
Chemical Reaction Engineering	10	Chapter 5
Chemical Engineering Thermodynamics	10	Chapter 3
Chemistry[1]	10	
Computer Usage in Chemical Engineering	5	Chapter 11
Fluid Dynamics[2]	10	
Heat Transfer	10	Chapter 7
Mass Transfer	10	Chapter 4
Material/Energy Balances	15	Chapter 2
Process Control	5	Chapter 9
Process Design and Economic Optimization	10	Chapter 6
Safety, Health, and Environmental	5	Chapter 12, Chapter 13

[1]See Chapter 14, *Fundamentals of Engineering: FE/EIT Exam Preparation*
[2]See Chapter 10, *Fundamentals of Engineering: FE/EIT Exam Preparation*

Some state boards administer the exam twice a year; others offer the exam only once a year. The scheduled exam dates for the next ten years can be found on the NCEES Web site (*www.ncees.org/exams/schedules/*).

Those wishing to take the exam must apply to their state board several months before the exam date.

Examination Procedure

Before the morning four-hour session begins, the proctors pass out exam booklets and a scoring sheet to each examinee. Space is provided on each page of the examination booklet for scratchwork. The scratchwork will *not* be considered in the scoring. Proctors will also provide each examinee with a mechanical pencil for use in recording answers; this is the only writing instrument allowed. Do not bring your own lead or eraser. If you need an additional pencil during the exam, a proctor will supply one.

The examination is closed book. You may not bring any reference materials with you to the exam. To replace your own materials, NCEES has prepared a *Fundamentals of Engineering (FE) Supplied-Reference Handbook.* The handbook contains engineering, scientific, and mathematical formulas and tables for use in the examination. Examinees will receive the handbook from their state registration board prior to the examination. The *FE Supplied-Reference Handbook* is also included in the exam materials distributed at the beginning of each four-hour exam period.

There are three versions (A, B, and C) of the exam. These have the major subjects presented in a different order to reduce the possibility of examinees copying from one another. The first subject on your exam, for example, might be fluid mechanics, while the exam of the person next to you may have electrical circuits as the first subject.

The afternoon session begins following a one-hour lunch break. The afternoon exam booklets will be distributed along with a scoring sheet. There will be 60 multiple choice questions, each of which carries twice the grading weight of the morning exam questions.

If you answer all questions more than 15 minutes early, you may turn in the exam materials and leave. If you finish in the last 15 minutes, however, you must remain to the end of the exam period to ensure a quiet environment for all those still working, and to ensure an orderly collection of materials.

Examination-Taking Suggestions

Those familiar with the psychology of examinations have several suggestions for examinees:

1. There are really two skills that examinees can develop and sharpen. One is the skill of illustrating one's knowledge. The other is the skill of familiarization with examination structure and procedure. The first can be enhanced by a systematic review of the subject matter. The second, exam-taking skills, can be improved by practice with sample problems—that is, problems that are presented in the exam format with similar content and level of difficulty.

2. Examinees should answer every problem, even if it is necessary to guess. There is no penalty for guessing.

3. Plan ahead with a strategy and a time allocation. A time plan gives you the confidence of being in control. Misallocation of time for the exam can be a serious mistake. There are 120 morning problems in 12 subject areas. Compute how much time you will allow for each of the 12 subject areas. You might allocate a little less time per problem for the areas in which you are most proficient, leaving a little more time in subjects that are more difficult for you. Your time plan should include a reserve block for especially difficult problems, for checking your scoring sheet, and finally for making last-minute guesses on problems you did not work. Your strategy might also include time allotments for two passes through the exam—the first to work all problems for which answers are obvious to you, the second to return to the more complex, time-consuming problems and the ones at which you might need to guess.

4. Read all four multiple-choice answers options before making a selection. All distractors (wrong answers) are designed to be plausible. Only one option will be the best answer.

5. Do not change an answer unless you are absolutely certain you have made a mistake. Your first reaction is likely to be correct.

6. If time permits, check your work.

7. Do not sit next to a friend, a window, or other potential distraction.

License Review Books

To prepare for the FE/EIT exam you need two or three review books.

1. A general review book for the morning exam, such as *Fundamentals of Engineering: FE/EIT Exam Preparation*, also from Kaplan AEC. That book will also prepare you for the general afternoon exam if you choose that option.

2. A review book for the afternoon exam, if you plan to take one of the discipline-specific exams.

3. *Fundamentals of Engineering (FE) Supplied-Reference Handbook*. At some point this NCEES-prepared book will be provided to applicants by their State Registration Board. You may want to obtain a copy sooner so you will have ample time to study it before the exam. You must, however, pay close attention to the *FE Supplied-Reference Handbook* and the notation used in it, because it is the only book you will have at the exam.

Textbooks

If you still have your university textbooks, they can be useful in preparing for the exam, unless they are out of date. To a great extent the books will be like old friends with familiar notation. You probably need both textbooks and license review books for efficient study and review.

Examination Day Preparations

The exam day will be a stressful and tiring one. You should take steps to eliminate the possibility of unpleasant surprises. If at all possible, visit the examination site ahead of time. Try too determine such items as

1. How much time should I allow for travel to the exam on that day? Plan to arrive about 15 minutes early. That way you will have ample time, but not too much time. Arriving too early, and mingling with others who are also anxious, can increase your anxiety and nervousness.

2. Where will I park?

3. How does the exam site look? Will I have ample workspace? Will it be overly bright (sunglasses), or cold (sweater), or noisy (earplugs)? Would a cushion make the chair more comfortable?

4. Where are the drinking fountains and lavatory facilities?

5. What about food? Most states do not allow food in the test room (exceptions for ADA). Should I take something along for energy in the exam? A light bag lunch during the break makes sense.

Items to Take to the Examination

Although you may not bring books to the exam, you should bring the following:

- *Calculator*—Beginning with the April 2004 exam, NCEES has implemented a more stringent policy regarding permitted calculators. For a list of permitted models, see the NCEES Web site (*www.ncees.org*). You also need to determine whether your state permits pre-programmed calculators. Bring extra batteries for your calculator just in case, and many people feel that bringing a second calculator is also a very good idea.

- *Clock*—You must have a time plan and a clock or wristwatch.

- *Exam Assignment Paperwork*—Take along the letter assigning you to the exam at the specified location to prove that you are the registered person. Also bring something with your name and picture (driver's license or identification card).

- *Items Suggested by Your Advance Visit*—If you visit the exam site, it will probably suggest an item or two that you need to add to your list.

- *Clothes*—Plan to wear comfortable clothes. You probably will do better if you are slightly cool, so it is wise to wear layered clothing.

Special Medical Condition

If you have a medical situation that may require special accommodation, you need to notify the licensing board well in advance of exam day.

Examination Scoring

The questions are machine-scored by scanning. The answer sheets are checked for errors by computer. Marking two answers to a question, for example, will be detected and no credit will be given.

Your state board will notify you whether you have passed or failed roughly three months after the exam. Candidates who do not pass the exam the first time may take it again. If you do not pass you will receive a report listing the percentages of questions you answered correctly for each topic area. This information can help focus the review efforts of candidates who need to retake the exam.

The FE/EIT exam is challenging, but analysis of previous pass rates shows that the majority of candidates do pass it the first time. By reviewing appropriate concepts and practicing with exam-style problems, you can be in that majority. Good luck!

About the Authors

Dilip K. Das, M.S., P.E., is a principal engineer in Bayer CropScience's Kansas City engineering department. He is currently in charge of emergency relief system design, and previously held process engineering positions at Ciba-Geigy, Rhone-Poulenc, and Stauffer Chemicals. He has a B.S.ChE (honors) from Jadavpur University, Calcutta, where he received gold medals for excellence, and an M.S.ChE from the University of Washington, Seattle. He also attended MIT for training in artificial intelligence in chemical engineering. He is the author of several technical articles and co-author of four books on chemical engineering, and holds a U.S. patent. Das is a registered engineer in the states of New York, New Jersey, Louisiana and Missouri. He is a past chair of AIChE's Kansas City chapter, and is currently the chairman of SuperChems Technical Steering Committee under the Design Institute of Emergency Relief Systems (DIERS). He has also been published at poetry.com.

Rajaram K. Prabhudesai, Ph.D., P.E., is currently a consulting chemical engineer. He worked as process supervisor for Badger Engineers where his responsibilities were process and plant design. Previously he was senior process engineer with Stauffer Chemical Company, where he was responsible for process design, systems engineering, process economics, and plant start-up. He also worked for AMF as principal research engineer and Coca Cola Company as principal chemical engineer. He has written numerous papers for professional journals and contributed to Perry's *Chemical Engineering Handbook* and Shweitzer's *Handbook of Separation Techniques for Chemical Engineer* (both McGraw-Hill). Dr. Prabhudesai received his M.S. from the University of Bombay and his Ph.D. in chemical engineering from the University of Oklahoma.

CHAPTER 1

Dimensions and Units

OUTLINE

UNITS OF FORCE 1

MOLAR UNITS 2

MULTIPLIERS AND CONVERSION FACTORS 2

REFERENCES 4

Dimensions are names used to describe the characteristics of a physical quantity. Examples of dimensions are length, mass, time, temperature, and electric charge. Units are the basic standards of measuring the magnitudes of physical quantities. For example, meter is a unit for the measurement of length, and second is a unit for measuring time. Units are two types: fundamental and derived. Fundamental units are independently defined, whereas derived units are expressed in terms of fundamental units. Various unit systems differ from one another in the choice of fundamental dimensions and units (Table 1.1).

UNITS OF FORCE

In the SI system, $\quad$ 1 Newton = (1 kg) × (1 m/s^2)

In the British Engineering system, $\quad$ $1 \text{ slug} = \dfrac{1 \text{ pound-force}}{1 \text{ ft/s}^2}$

In the fps system, $\quad$ 1 poundal = (1 lb$_m$) × (1 ft/s^2)

Table 1.1 Dimensions and Units

Dimension	Symbol	Metric or SI System Unit	British Unit	USCS Unit
Length	L	meter	foot	foot
Mass	M	kg	slug*	lb$_m$
Time	t	second	second	second
Temperature	T	degree K	°F	°F
Force	F*	Newton (N)*	poundal	lb$_f$

*Denotes a derived unit in the system.

Comparison of British and fps systems shows the following:

$$1 \text{ slug} = 32.174 \text{ lb}_m$$
$$1 \text{ lb-force (lb}_f) = 32.174 \text{ poundals}$$

The US customary system, $1 \text{ lb-force (lb}_f) = \dfrac{(1 \text{ lb}_m)(1 \text{ ft/s}^2)}{g_c}$

Where Newton's law proportionality factor $g_c = 32.174 \text{ lb}_m \cdot \text{ft/s}^2 \cdot \text{lb}_f$

MOLAR UNITS

A mole of a substance equals the molecular mass expressed in given units. Molecular mass is also commonly called molecular weight, although weight implies force.

$$1 \text{ kilogram-mole} = \text{molecular weight in kilograms}$$
$$1 \text{ pound-mole} = \text{molecular weight in pounds}$$
$$1 \text{ gmol} = \text{molecular weight expressed in grams}$$
$$1 \text{ kmol} = 1000 \text{ gmol}$$

The mole is the SI unit for amount of substance. One mole contains as many entities as the number of atoms in exactly 12 grams of carbon-12. Its symbol is mol. One atomic mass unit (amu) is one-twelfth of the mass of one ^{12}C atom = 1.661×10^{-24} g.

The number of atoms in 12 grams of ^{12}C is 6.0221367×10^{23}. Each molecule has a mass of MM amu where MM = molecular mass. Thus 1 mol of water contains 6.0221367×10^{23} molecules. Therefore, the mass of 1 mol of water is given by

$$m = 6.0221367 \times 10^{23} \text{ molecules} \left(\frac{18 \text{ amu}}{\text{molecule}} \right) \left(\frac{1.661 \times 10^{-24} \text{ g}}{\text{amu}} \right) = 18 \text{ g}$$

MULTIPLIERS AND CONVERSION FACTORS

For practical purposes, multiples or submultiples of the same unit are used and are given special names. For example, 1 ft = 12 inches. Here an inch is a unit, which has a value of length equal to 1/12 of one foot. The metric multiplier factors are given in Table 1.2 and some basic conversion factors are given in Table 1.3.

In practice, one uses conversion factors already tabulated in various references.[1-4] Some factors that are required very often are given in Table 1.4.

Table 1.2 Multipliers of Standards Units SI or Metric System

Multiplier	Prefix of Standard Unit	Symbol	Multiplier	Prefix of Standard Unit	Symbol
10^9	giga	G	10^{-9}	nano	n
10^6	mega	M	10^{-6}	micro	μ
10^3	kilo	k	10^{-3}	milli	m
10^2	hecto	h	10^{-2}	centi	c
10^1	deka	da	10^{-1}	deci	d

Table 1.3 Some Basic Conversion Factors

Length:	1 meter = 3.281 ft	1 inch = 2.54 cm = 25.4 mm
	1 mile = 5280 ft = 1.609 km	1 foot = 30.48 cm
	1 foot = 12 inches	
Mass:	1 kg = 2.2046 lb_m	1 lb_m = 453.6 gm = 0.4536 kg
	1 U.S. ton = 2000 lb_m	1 ton (British) = 2240 lb_m
	1 metric ton = 1000 kg	1 slug = 32.174 lb_m
Time:	1 second = (1/3600) h = (1/60) minute	
Force:	1 lb_f = 32.174 poundals = 4.448 N	
	1 Newton = 10^5 dyne	1 poundal = 0.138 newton
Temperature:	1 Kelvin = °C + 273.16, °R = °F + 459.7	
Pressure:	1 bar = 1 × 10^5 Pa	1 atm = 1.013 Bar = 1.03 kg/cm^2
	1 Pa = 1 N/m^2	1 atm = 760 mm Hg
Thermal units:	1 calorie = 4.1868 J	1 kcal = 4186.8 J
	1 Btu = 0.252 kcal = 778.2 ft·lb_f	

Table 1.4 Conversion Factors

Multiply	By	To Obtain
Atm std	14.7	psi
Atm std	760	mm Hg
Atm std	29.92	in Hg
Atm std	33.9	ft of water
Atm std	1.013 × 10^5	Pa
Atm std	1.013	Bar
Atm std	1.033	kg/cm^2
Atm std	101.3	kPa
Bar	1 × 10^5	Pa
Bar	1.0197	kg/cm^2
Bar	14.507	psi
Btu	1054.4	Joule(j)
Btu	2.928 × 10^{-4}	kWh
Btu	778.2	ft·lb_f
Btu	0.252	kcal
Btu/h	3.93 × 10^{-4}	hp
Btu/h	0.293	watt(W)
Btu/h	0.22	ft·lb_f/s
Btu/h·ft^2·°F	4.88	kcal/h·m^2·°C
Btu/h·ft^2·°F	5.67 × 10^{-3}	kW/m^2·°C
Btu/h·ft^2·°F	5.68	J/(s·m^2·°C)
Btu/(h·ft^2·°F/ft)	1.49	kcal/h·°C·m
Btu/(h·ft^2·°F/ft)	1.73	J/s·m·°C
Btu/lb·°F	1	kcal/kg·°C
cP	1 × 10^{-3}	N·s/m^2
cP	2.42	lb/h·ft
cP	0.000672	lb/s·ft

(Continued)

Table 1.4 Conversion Factors (Continued)

Multiply	By	To Obtain
cP	3.6	kg/h·m
cP	1	mPa·s
ft^3	7.48	gal
Ft·lb$_f$	1.285×10^{-3}	Btu
Ft·lb$_f$	3.766×10^{-7}	kWh
Ft·lb$_f$	0.32	cal
Ft·lb$_f$	1.36	joule(J)
Ft·lb$_f$/s	1.818×10^{-3}	hp
gal	3.785	liters(L)
hp	33,000	Ft·lb$_f$/min
hp	42.4	Btu/min
hp	745.7	watt(W)
Hp·h	2545	Btu
Hp·h	2.68×10^6	joule(J)
joule(J)	9.478×10^{-4}	Btu
joule(J)	0.74	Ft·lb$_f$
joule(J)	1	N·m
joule(J)/s	1	watt(W)
J/(s·m^2·°C)	1	W/m^2·°C
kcal	3.9685	Btu
kcal	1.56×10^{-3}	Hp·h
kcal	4.186	joule(J)
kN/m^2	0.295	in Hg
kPa	0.15	psi
kW	1.34	hp
kW	737.6	ft·lb$_f$/s
kWh	3413	Btu
kWh	3.6×10^6	joule(J)
kip	1000	lb$_f$
mile	1.609	km
micron	1×10^{-6}	meter
poundal	0.14	Newton
watts	3.41	Btu/h
watts	1	Joule/s
W/m^2	0.317	Btu/h·ft^2
W/m^2·K	0.1761	Btu/h·ft^2·°F
W/(m^2·K/m)	0.58	Btu/(h·ft^2·°F/ft)

REFERENCES

1. Das, D. K., and R. K. Prabhudesai. *Chemical Engineering License Review*, 2nd ed., Kaplan AEC Education, 2004.
2. National Council of Examiners for Engineering and Surveying. *Fundamentals of Engineering Supplied-Reference Handbook*, 6th ed., NCEES 2003.
3. Noble, R. D. Material and Energy Balances, *AIChE Modular Instruction,* vol. 1, *Series F,* 1981.
4. Perry R. H. *Chemical Engineers' Handbook,* Platinum ed. McGraw-Hill Book Co., 2000.

PROBLEMS

1.1 Choose the correct answer for each of the following questions.
   (a) In addition to length, time, and pound-force, mass is also a fundamental dimension in the British engineering system.
       (1) True
       (2) False

   (b) Dimensional consistency of an equation guarantees its accuracy.
       (1) True
       (2) False

   (c) Conversion factors when utilized are dimensionless.
       (1) True
       (2) False

   (d) Candela is the fundamental unit for luminous intensity in the SI system.
       (1) True
       (2) False

   (e) The following equation relating the area of a right circular cone of altitude h and base radius of r is dimensionally consistent.
       $$a = \pi r^2 + \pi r (r^2 + h^2)^{1/2}$$
       (1) True
       (2) False

   (f) Specific weight of a substance is defined as its weight per unit volume. When the SI system of units is used, the specific weight w of a substance is given by
       $$w = \rho g.$$
       (1) True
       (2) False

   (g) A stone having a mass of 10 kg, is moving with a velocity of 3 m/s. Its kinetic energy is therefore 0.045 kJ.
       (1) True
       (2) False

   (h) The pressure gauge on a pipeline in a plant reads 4 bar. The barometer reads 752.3 mm Hg. The pipeline pressure is therefore 5 bara.
       (1) True
       (2) False

   (i) Water is fed to a steam boiler at a rate of 455 m^3/h. Water analysis shows its oxygen content to be 8 ppm. The total dissolved oxygen fed with the water is therefore 3.64 kg/h.
       (1) True
       (2) False

(j) Heat input into a vessel due to external fire is given by

$$Q = 21000 (A)^{0.82}$$

wherein Q is in Btu/h and A is wetted area of the vessel in ft². The heat input in kW is therefore given by

$$Q_m = 43.19(A_m)^{0.82}$$

Wherein Q_m = heat input in kW and A_m = wetted area of vessel in m².
(1) True
(2) False

1.2 Using the fundamental definitions and units, calculate the conversion factor for 1 std atm to bar given the following data: density of Hg = 13.595 g/cm³, g = 9.807 m/s².

1.3 An equation for the molar heat capacity, C_p, of methane gas at constant pressure and over a temperature range of 273–1200 K (Perry's Handbook, 6th ed., pp. 3–130) is given by:

$$C_p = 5.34 + 0.0115T$$

where
C_p is in cal/(gmol·K)
T is in degrees Kelvin

Obtain an expression for the heat capacity when specific heat is desired in units of kJ/(kg·°C) and the temperature in °C.

SOLUTIONS

1.1 a. False.
The British Engineering system uses time, length, and force as fundamental units; mass is a derived unit in this system.

b. False.
Dimensional inconsistency of an equation shows that it is wrong, but the consistency is not enough to prove its accuracy because in the analysis performed all the variables involved may not have been accounted for.

c. True.
The operation of using dimensionless conversion factors is equivalent to multiplication by unity. The conversion factor is a numerical ratio of two numbers and is dimensionless. One number is in one set of units, whereas the other is in a different set of units.

d. True.
Its symbol is cd.

e. True.
The dimensions on the left of the equation are L^2, and the same is true for the right hand side.

f. True.
Using Newton's second law, $F = ma$, and dividing both sides of the force equation by V (volume of the substance), one obtains $F/V = (m/V)\,a$, which can be written as $w = \rho g$ when $a = g$, where ρ is the density of the substance in kg/m^3 and g is the local acceleration of gravity.

g. True.
$$KE = mu^2/2 = (10 \text{ kg})(3^2 m^2/s^2)/2 = 45 \text{ kg-}m^2/s^2 = 0.045 \text{ kJ}$$

h. True.
The pressure gauge indicates gage pressure. By adding local barometric pressure = $(750.3/760) \times 1.013 = 1$ bar to it, the pressure 5 bar in absolute units is obtained.

i. True.
ppm means one part in one million by weight. 8 ppm means 8 parts per million by weight. This is equivalent to a weight fraction of 0.000008. ρ of water 1000 kg/m^3. Hence oxygen in water = 455 (1000) × 0.000008 = 3.64 kg/h.

j. True.
$[1\text{m}^2 = (3.281)^2 \text{ft}^2]$ and $1 \text{ Btu/h} = 0.293 \times 10^{-3}$ kW. When these conversion factors are substituted in the equation using British units and numerical values are combined, the constant in SI units results as follows:

$$A(\text{ft}^2) = [(3.281)^2 (\text{ft}^2/\text{m}^2)] A_m (\text{m}^2) = 10.765 A_m$$

Then

$$Q_m = 21{,}000(10.765 A_m)^{0.82} \left(\frac{\text{Btu}}{\text{h}}\right) \times \left(0.293 \times 10^{-3} \frac{\text{kW}}{\text{Btu/h}}\right)$$

$$= 43.19(A_m)^{0.82} \text{ kW}$$

1.2 $\quad$ **1 std atm** $= \rho g \Delta z = 13.595 \dfrac{\text{g}}{\text{cm}^3} \times 9.807 \text{ m/s}^2 \times 760 \text{ mmHg}$

$$= \frac{13.595}{1000 \frac{\text{g}}{\text{kg}}} \frac{\text{g}}{\text{cm}^3 \times \frac{10^{-6} \text{ m}^3}{\text{cm}^3}} \times 9.807 \text{ m/s}^2 \times 760 \text{ mm} \times \frac{10^{-3}}{\text{mm}}$$

$$= 101{,}328 \frac{\text{kg}}{\text{m} \cdot \text{s}^2}$$

$$= 101{,}328 \frac{\text{kg}}{\text{m} \cdot \text{s}^2} \left(\frac{1 \text{N} \cdot \text{s}^2}{1 \text{kg} \cdot \text{m}}\right) = 101{,}328 \left(\frac{\text{N}}{\text{m}^2}\right)$$

$$= 101{,}328 \text{ Pa} = 101.3 \text{ kPa}$$

$$= 101.3/100 = 1.013 \text{ bar}$$

1.3 $\quad$ Molecular weight of $CH_4 = 16$

1 g of $CH_4 = (1/16)$ gmol CH_4
1 cal = 4.184 J; 1°C = 1 K;
$T = t_c + 273.16$ °C

Given C_p = Molar heat capacity, $\dfrac{\text{cal}}{\text{gmol} \cdot \text{K}}$, Required c_p = heat capacity in $\dfrac{\text{kJ}}{\text{kg} \cdot °\text{C}}$

$$C_p = (5.34 + 0.0115T) \, \dfrac{\text{cal}}{\text{gmol} \cdot \text{K}}$$

$$= [5.34 + 0.0115(t_c + 273.16)] \, \dfrac{\text{cal}}{\text{gmol} \cdot \text{K}} \, \dfrac{4.184 \text{ J/cal}}{(16 \text{ g/gmol}) \cdot °\text{C}}$$

$$= (0.2615)[8.4813 + .0115 t_c] \, \dfrac{\text{J}}{\text{g} \cdot °\text{C}}$$

$$= [2.2179 + 3.0073 t_c] \, \dfrac{\text{J}}{\text{g} \cdot °\text{C}}$$

Therefore, heat capacity $c_p = (2.2179 + 3.0073 t_c) \, \dfrac{1000 \text{ J}}{1000 \text{ g} \cdot °\text{C}}$

$$= (2.2179 + 3.0073 t_c) \, \dfrac{\text{kJ}}{\text{kg} \cdot °\text{C}}$$

Where t_c is in °C.

CHAPTER 2

Material and Energy Balances

OUTLINE

ACCOUNTING PRINCIPLE 12

DEFINITIONS AND RELATIONSHIPS OF MATERIAL PROPERTIES 12

VAPOR PRESSURE 13

IDEAL GASES 14

IDEAL GAS MIXTURES 15

CRITICAL PROPERTIES 15
Corresponding State

NON-IDEAL OR REAL GASES 15
Equations of State for Non-Ideal (Real) Gases

SOLUTIONS 16
Relative Vapor Pressure ■ Boiling Point Elevation ■ Freezing Point Depression

HUMIDITY AND SATURATION 17
Water Vapor-Gas Mixtures

SOLUBILITY AND CRYSTALLIZATION 18

MASS BALANCES 18

ENERGY BALANCES 19

LAW OF THERMOCHEMISTRY 21

FUELS AND COMBUSTION 21

REFERENCES 22

Chapter 2 Material and Energy Balances

Industrial process analysis involves application of the principles of chemistry and physics and requires understanding of the properties of materials, behavior of gases and liquids under different conditions, and mass and energy balances. This chapter includes a review of the following topics: properties of materials, process stoichiometry, laws of conservation of mass and energy, and thermochemistry.

ACCOUNTING PRINCIPLE

The accounting principle can be expressed by the equation:

$$Q_E - Q_B = \Sigma Q_I - \Sigma Q_O + \Sigma Q_P$$

where
- Q = the physical quantity chosen for accounting
- Q_E = the quantity at the end of the accounting period
- Q_B = the quantity at the beginning of the accounting period
- Q_I = the quantity that enters the system during the accounting period
- Q_O = the quantity that leaves the system during the accounting period
- Q_P = the quantity created or destroyed during the accounting period

In ordinary processes involving chemical (not nuclear) reactions, $\Sigma Q_P = 0$. If the quantity in the system does not change with time, $Q_E - Q_B = 0$ and the accounting equation reduces to $\Sigma Q_I - \Sigma Q_O = 0$. In this case, input equals output. The system is said to be under *steady state condition*. The accounting principle can also be used to make an energy balance for a system.

DEFINITIONS AND RELATIONSHIPS OF MATERIAL PROPERTIES

In carrying out mass and energy balances, many other derived quantities defined in terms of the fundamental units are used because of their convenience. This section briefly defines several such quantities.

Density of a substance, ρ = mass/volume

Specific gravity of a substance, sp gr = ρ/ρ_r

where, ρ_r is the density of the reference substance; for liquids and solids, water is usually the reference substance at 4°C ($\rho = 1000$ kg/m^3)

Other specific gravity conventions are in use. For example in the petroleum industry the API gravity scale is used and is defined as °API = (141.5/sp gr) − 131.5. Other specific gravity scales include Baumè, Brix, and Twadell.

Specific volume = $V = 1/\rho$

Mass fraction = (mass of component A)/(Total mass of all components present)

Mass % = mass fraction × 1000

Mol fraction x_A = (Moles of component A)/(Total moles of all components present)

where
x_A = mol fraction of component

Mol fraction in vapor phase is denoted by y, and in liquid it is usually denoted by x.

Mol % = mole fraction × 100.

Concentration of a species in a mixture is defined as its quantity per unit volume of the mixture.

Volume concentration of $A = V_a/V_M$

where
V_a = Volume of component A
V_M = Volume of mixture
V_A and V_M are at the temperature and pressure of the mixture

Mass concentration of a species in a mixture = (Mass of species i)/Volume of mixture

Average molecular weight of a gas mixture $M_{av} = \sum x_i M_i$

where
x_i and M_i are the mol fraction and molecular weight of component i

Molar concentration of a species i in a mixture = (Moles of species i)/Volume of mixture

Molarity = M = (gmol of species i in a mixture)/1 liter of solution

Normality = N = (gram-equivalent weight of species i)/1 liter of solution

Molality = m = gmol of species i/kg solvent

Flow rate is the amount of material that passes a given reference point in a system per unit of time

Mass flow rate and volumetric flow rate are the most important methods of expressing flow rates

VAPOR PRESSURE

The pressure exerted by a vapor in equilibrium with its liquid at a given temperature is termed its vapor pressure. If the gas phase consists of more than one component, the total pressure exerted on the liquid surface is the sum of the pressures exerted by the various components. The pressure contribution by a component to the total pressure is termed its partial pressure.

At saturation, component partial pressure equals component vapor pressure. The temperature at which a vapor is saturated is called the *dew point or saturation temperature.*

The *boiling point* of a liquid at a given pressure is defined as the temperature at which its equilibrium vapor pressure equals the total pressure on its surface. The temperature at which a liquid boils under a total pressure of 1 atm is called its *normal boiling point.*

The *vapor pressure* of a pure substance is a unique property of that substance and increases with temperature. Two frequently used relations for the vapor pressure are as follows:

Two constant equation

$$\ln p = A + B/T$$

where
- p = vapor pressure
- T = absolute temperature, K

Antoine equation

$$\ln p = A - [B/(T + C)]$$

where
- A, B, and C are constants
- T is in degrees K

An important equation relating vapor pressure, heat of vaporization, and temperature is the Clausius-Clapeyron equation:

$$\frac{dp}{p} = \frac{\Delta H_v \, dT}{RT^2}$$

where
- R = Gas-law constant
- ΔH_v = heat of vaporization

Watson's empirical equation correlating heat of vaporization and temperature is

$$\frac{\lambda_2}{\lambda_1} = \left(\frac{1 - T_{r2}}{1 - T_{r1}}\right)^{0.38}$$

where
- λ_2 = Heat of vaporization cal/gmol at temperature T_2
- λ_1 = Heat of vaporization cal/gmol at temperature T_1
- T_{r2} = Reduced temperature at T_2 and T_{r1} = Reduced temperature at T_1

IDEAL GASES

Boyle's Law PV = Constant at constant temperature

Charles's Law P/T = Constant at constant volume

Equation of state $PV = nRT$ where R is gas law constant and n = number of moles

It should be particularly noted that in the preceding equations the temperature is in K or °R, and P is the absolute (not the gage) pressure in appropriate units. The values of gas constant R will depend upon the units chosen for P, V, and T.

IDEAL GAS MIXTURES

Dalton's Law: $P = p_A + p_B + p_C + \cdots$

Amagat's Law: $V = V_A + V_B + V_C + \cdots$

$$p_A = y_A P$$
$$V_A = y_A V$$

Ideal gas law is applicable at low pressure and high temperatures

Normal molar volume = 22.4 m³/kg mol at 0°C and 1.0 atm (760 mm Hg)

Normal molar volume = 22.4 liters/gmol at 0°C and 1.0 atm (760 mm Hg)

Normal molar volume = 359 ft³/lbmol of an ideal gas at 32°F and 1 atm

1 lbmol of an ideal gas occupies a volume of 379 ft³ at 60°F and 1 atm pressure

CRITICAL PROPERTIES

Critical temperature, T_c—Temperature at which the molecular kinetic energy of translation equals the maximum potential energy of attraction

Critical pressure, P_c—Pressure required to liquefy a gas at its critical temperature

Critical volume, V_c—Volume of gas at critical state (at P_c and T_c)

Reduced Conditions—Reduced temperature, $T_r = T/T_c$

Reduced pressure, $P_r = P/P_c$

Reduced volume, $V_r = V/V_c$

Corresponding State

This is the state of equal reduced temperature and pressure. Many properties of gases and liquids are nearly the same at corresponding state.

NON-IDEAL OR REAL GASES

For real gases a compressibility factor is defined by the following equation

Compressibility of an actual gas, $z = PV/RT$

Critical compressibility is the compressibility at the critical point and is given by the relation

$$z_c = P_c V_c / RT_c$$

Equations of State for Non-Ideal (Real) Gases

Many relations have been proposed to represent *PVT* data by a single equation so that *PVT* data for all materials can be calculated by assigning appropriate values to the constants in the equation. Some commonly used equations are

- **Van der Waals' Equation:** $(P + a/v^2)(v - b) = RT$

where
- $n = 1$
- $v_c = 3b$
- $T_c = 8a/27Rb$
- $P_c = a/27b^2$

- **Redlich-Kwong Equation:** $P = RT/(v - b) - a/[T^{0.5} v(v + b)]$

where
- $a = 0.42748\ R^2 T_c^{2.5}/P_c$
- $b = 0.08664\ RT_c/P_c$

- **Virial equations** correlating *PVT* data of real gases are power series either in *P* or 1/*V*. The constants in these equations are known as virial coefficients. These equations reduce to *PV/RT* at low pressures. One form of virial equation is

$$\frac{Pv}{RT} = 1 + \frac{B(T)}{v} + \frac{C(T)}{v^2} + \frac{D(T)}{v^3} + \cdots$$

The Benedict-Webb-Ruben equation with eight constants and the Beattie-Bridgeman equation with five constants are the most important virial equations.

Another generalized equation, which retains ideal gas law appearance, is given by

$$PV = znRT$$

where
z is the compressibility factor as defined earlier and is a function of the pressure and temperature

Additional discussion of *PVT* data calculations is found in Chapter 3.

SOLUTIONS

For substances in dilute solutions, Raoult's law applies. It states that the equilibrium vapor pressure exerted by a component in a solution is proportional to its mole fraction in the solution. Thus

$$p_A = x_A P_A$$

where
- p_A = vapor pressure of component *A* in solution with components *B*, *C*
- x_A = mole fraction of *A* in solution
- P_A = vapor pressure of pure component *A* at the temperature of the solution

Relative Vapor Pressure

A modified form of Raoult's law is often used to obtain vapor pressure data for a component from a single experimental point. This is given by

$$p = kp_s$$

where
- p = vapor pressure of solution
- p_s = vapor pressure of pure solvent
- k = a proportionality factor called the relative vapor pressure of the solution

Boiling Point Elevation

The boiling point elevation is the increase in boiling point of a solution compared to pure solvent due to the presence of a non-volatile solute in the solvent. The equation is

$$\text{B.P. elevation} = \frac{RT^2}{\Delta H_v} \frac{M_s}{1000} \times \frac{w}{M_a}$$

where
- T = boiling point of pure solvent, K
- M_s = molecular weight of solvent
- M_a = molecular weight of solute
- ΔH_v = Molar heat of vaporization of the solvent, cal/gmol
- w = grams of solute in 1000 grams of solvent

Freezing Point Depression

The freezing point depression refers to the lowering of the freezing point of a solution compared to pure solvent due to the presence of a nonvolatile solute. The equation is

$$\text{Freezing point depression} = \frac{RT^2}{\Delta H_f} \frac{M_s}{1000} \times \frac{w}{M_a}$$

where
- ΔH_f = latent heat of fusion, cal/gmol

HUMIDITY AND SATURATION

Pure component volume in a saturated gas obeying ideal gas law is given by

$$V_v = V(p_v/p)$$

where
- V_v = pure component volume of vapor
- p_v = partial pressure of vapor
- p = total pressure

For unsaturated gas mixtures, partial pressure of vapor is less than the vapor pressure.

Relative saturation in % = $(p_v/p_s) \times 100 = y_r$

Percent saturation = $(n_v/n_s) \times 100 = y_p$

where
n_v = moles of vapor per mole of vapor-free gas actually present
n_s = moles of vapor per mole of vapor-free gas when the mixture is saturated

$$\text{Applying Dalton's Law,} \quad y_p = y_r \left(\frac{p - ps}{p - pv} \right)$$

Water Vapor-Gas Mixtures

Humidity, H = weight of water vapor per unit weight of moisture-free gas

Molal humidity, H_m = number of moles of water vapor per mole of moisture-free gas

% Humidity = % saturation when applied to water

Wet-bulb temperature is the equilibrium temperature attained by a liquid vaporizing into a gas under adiabatic conditions

Dry-bulb temperature is the temperature of the gas into which evaporation takes place

Psychrometry is the application of wet and dry-bulb thermometry to air-water system. Humidity chart for water is constructed usually for a total pressure of 1 atm.

SOLUBILITY AND CRYSTALLIZATION

Solubility is the concentration of the solute in a saturated solution at the temperature of the solution. Solutes are of two types: those that form compounds with the solvent and those that do not. *Solvates* are compounds of definite proportions between solute and solvent. If water is the solvent, the compound is termed *hydrate*.

Solubility diagrams are available for a number of solute-solvent pairs in the literature. Material balance problems involving dissolution of solute in a solvent or its crystallization from the solvent can be solved using solubility diagrams.

In solvent extraction, the initial solvent from which a soluble component is to be removed is called *raffinate solvent*, and the immiscible solvent used for extracting the solute is termed *extract solvent*. If equilibrium exists between the raffinate and the extract phases, the distribution coefficient for the solute is given by

$$K = \frac{C_E}{C_R}$$

where
  K = distribution coefficient
  C_E = solute concentration in extract phase
  C_R = solute concentration in raffinate phase

MASS BALANCES

One of the most important applications of the general accounting principle in process analysis is the mass balance of a system over a chosen accounting period. The mass balance equation in its most general form for a given system can be directly written from the accounting principle as follows

$$M_E - M_B = \sum M_I - \sum M_O + \sum M_p - \sum M_C$$

where
- M_E = mass of the system at the end of the accounting period
- M_B = mass of the system at the beginning of the accounting period
- $\sum M_I$ = sum of all masses that entered the system during the accounting period
- $\sum M_O$ = sum of all masses that left the system during the accounting period
- $\sum M_p$ = mass changes resulting from physical or chemical transformations
- $\sum M_C$ = mass changes connected with atomic transmutations and relativistic effects; this term is zero in ordinary chemical and physical transformations

When applying the mass balance equation to a given system under study, the following points are worth noting:

(1) If the system undergoes only physical changes, $\sum M_p = 0$ and $\sum M_C = 0$.

(2) If the system undergoes a chemical transformation and no atomic transmutation, $\sum M_C = 0$ and $\sum M_p$ may be zero or non-zero depending on the system chosen. For the system as a whole $\sum M_p = 0$ (law of conservation of mass applies if the mass balance is considered). If one of the component species is chosen as a system, $\sum M_p$ is not equal to zero.

In ordinary physical and chemical processes, there are no atomic transmutations or relativistic effects and, consequently, $\sum M_C = 0$. Therefore, the mass balance equation for such processes reduces to

$$M_E - M_B = \sum M_I - \sum M_O + \sum M_p$$

This is the most common form of the mass balance equation in carrying out mass balances for physical and chemical processes encountered in process analysis.

(3) If a mole balance is considered, moles may not be conserved.

(4) For a steady state system, $M_E = M_B$, which means mass within the confines of the system remains constant.

ENERGY BALANCES

Closely related to the mass balance equation is the energy balance equation for a system. The energy balance equation can also be written using the general accounting principle. However, definitions of some energy terms are needed before the energy balance for a system is written down. The energy terms are

Potential energy, PE—external energy possessed by a system due to its position in a gravitational field

Kinetic energy, KE—all energy of a system associated with the macroscopic motion of its mass = $M(u^2/2g_c)$

Internal energy, U—total energy possessed by a system due to its component molecules, their relative positions, and their movement; for a closed system (constant mass), it is a state property

Heat energy, Q—energy transferred across the boundaries of a system because of the temperature difference

Work, W—work is defined as W = force × distance; it is a form of energy transferred across the boundaries of a system and is not system state property

Enthalpy, H—defined as $H = U + PV$

Both U and H are determined relative to a selected reference state. The reference state of zero enthalpy is chosen as 1 atm and 0 °C. For water, it is taken as 0 °C and its own vapor pressure at 0 °C.

The accounting principle can now be applied to energy balances of physical and chemical processes in which atomic transmutations and relativistic effects are absent. The equation can be written as follows:

$$(U + PE + KE)_E - (U + PE + KE)_B = \Sigma(H + PE + KE)_I - \Sigma(H + PE + KE)_O + Q + W$$

where

$(U + PE + KE)_E$ = sum of internal, potential, and kinetic energies at the end of accounting period

$(U + PE + KE)_B$ = sum of internal, potential, and kinetic energies at the beginning of accounting period

$(H + PE + KE)_I$ = energy transferred into the system that is associated with mass transfer

$(H + PE + KE)_O$ = energy transferred out of the system that is associated with mass transfer

Q = heat transferred across system boundaries during accounting period

W = all work that crosses system boundaries during the accounting period

Heat and work are energies in transit across the boundary of the system. They are not system properties, nor are they state properties as their quantities depend upon the path taken.

For a flow reaction with negligible changes in potential and kinetic energies and with no work, the energy balance reduces to $Q = \Delta H$. For a non-flow reaction taking place at constant pressure, $Q = \Delta H$. For a non-flow reaction taking place at constant volume, the energy balance reduces to $Q = \Delta U$.

Several key terms with respect to enthalpy changes that take place in chemical reactions are

Heat of chemical reaction is the change in enthalpy of the reaction system at constant pressure. This is a function of the nature of reactants and products involved and also their physical states.

Standard heat of reaction is the change of enthalpy when the reaction takes place at 1 atm pressure in such a manner that it starts and ends with all involved materials at a constant temperature of 25 °C. In exothermic reactions, heat is evolved. In endothermic reactions, heat is absorbed.

Heat of formation of a chemical compound is the change in enthalpy when elements combine to form the compound, which is the only reaction product. Standard heat of formation of a compound, ΔH_f, is the heat of reaction at 25 °C.

LAW OF THERMOCHEMISTRY

At a given temperature and pressure, the heat of formation of a compound from its elements is equal to the heat required to decompose the compound into its elements. The total change in enthalpy of a system depends only on the temperature, pressure, state of aggregation, and state of combination at the beginning and at the end of the reaction and is independent of the number of intermediate chemical reactions involved. This principle is also known as the Law of Hess. This principle is used to calculate the heats of formation of compounds knowing the heats of intermediate reactions. For example, the following three reactions enable one to obtain the heat of formation of ethane.

$$C_2H_6(g) + 3\tfrac{1}{2}O_2(g) = 2CO_2(g) + 3H_2O(l) \quad \Delta H_1 = -372.82 \text{ kcal/gmol} \quad \text{(a)}$$

$$C(\beta) + 2O_2(g) = 2CO_2(g) \quad \Delta H_2 = -188.1 \text{ kcal/gmol} \quad \text{(b)}$$

$$3H_2(g) + 1\tfrac{1}{2}O_2(g) = 3H_2O(l) \quad \Delta H_3 = -204.95 \text{ kcal/gmol} \quad \text{(c)}$$

Equation(b) + Equation(c) − Equation(a) gives

$$2C(\beta) + 3H_2(g) = C_2H_6(g)$$

and ΔH_f of ethane = −188.1 − 204.95 − (−372.82) = −20.23 kcal/gmol

Standard heat of combustion, ΔH_c, is the enthalpy change resulting from the combustion of a substance in its normal state at 25 °C and atmospheric pressure, with the combustion beginning and ending at 25 °C. Generally, gaseous CO_2 and liquid water are the combustion products. It is possible to calculate the heat of formation of a compound from its heat of combustion provided the heats of formation of other substances that participate in the reaction are known.

Standard heat of reaction, ΔH_R, can be calculated from the heats of formation of reactants and the products using the following relation

$$\left[\Delta H_{\text{reaction}} = \Sigma \Delta H_{f(\text{products})} - \Sigma \Delta H_{f(\text{reactants})}\right] \text{ at } 25°C$$

It can also be calculated from the heats of combustion of the reactants and products by the relation

$$\left[\Delta H_R = \Sigma \Delta H_{c(\text{reactants})} - \Sigma \Delta H_{c(\text{products})}\right] \text{ at } 25°C$$

(For additional review of energy balances, see Chapter 3.)

FUELS AND COMBUSTION

The heating value of fuel is its standard heat of combustion numerically but with the opposite sign. Two types of heating values are defined. *Total heating value* is the heat evolved in the complete combustion of a fuel under constant pressure at temperature of 25 °C with all the water initially present and that produced in the combustion condensed to the liquid state at 25 °C. This is also referred to as higher or gross heating value and is denoted as *HHV*. *Net heating value*, also termed *LHV* or lower heating value, is defined in a similar manner but the final state of the water is taken as vapor at 25 °C.

Ultimate analysis of coal involves the determination of all elements in the sample, for example, carbon, hydrogen, sulfur, and so on. *Proximate analysis* of coal involves determination of four groups, namely, fixed carbon, volatile matter, ash, and moisture. Total heating value of coal in Btu/lb is

$$\text{H.V.} = 14490C + 61000H_a + 5550S \quad \text{(This is Dulong's formula)}$$

where
  C = weight fraction of carbon
  H_a = weight fraction of hydrogen
  S = weight fraction of sulfur

$$\text{Net H.V. of coal} = \text{Total H.V.} - 8.94 \times H \times 1050$$

where
  H = weight fraction of total hydrogen, including available hydrogen, the hydrogen in moisture, and hydrogen in combined water

For petroleum fuels, the U.O.P. characterization factor is given by

$$K = \frac{\sqrt[3]{T_B}}{sp\,gr}$$

where
  K = U.O.P. characterization factor
  T_B = average boiling point, °R at 1 atm
  $sp\,gr$ = specific gravity at 60°F

REFERENCES

1. Das, D. K. and R. K. Prabhudesai. *Chemical Engineering License Review*, 2nd ed., Kaplan AEC Education, 2004.
2. Hougen O. A. et al. *Chemical Process Principles, Part 1*, New York, John Wiley & Sons, 1954.
3. Henley E. J., et al. *Material and Energy Balance Computations*, New York, John Wiley & Sons, 1969.
4. *AIChE Modular Instruction Series*, Material & Energy Balances, vol 1–4.
5. Perry R. H. and D. W. Green. *Chemical Engineers' Handbook*, Platinum ed., New York, McGraw-Hill 1999.

PROBLEMS

2.1 A fuel gas has the following volumetric composition. Assume ideal gas behavior.

Methane 85%, Ethane 10.5%, Nitrogen 4.5%
Calculate
(a) Composition of the gas in mole %
(b) Composition in wt%
(c) Average molecular weight of the gas
(d) Density of gas at normal conditions in (1) kg/m^3 (2) lb/ft^3
(e) Partial pressure of nitrogen in kPa if the total pressure is 1.013 barG, and barometric pressure is 1.013 barA.

2.2 2000 kg/h of a mixture consisting of 60 wt% benzene and 40 wt% toluene is to be separated in a distillation column. The distillate is to contain 98 wt% benzene and 95% of feed benzene is to be recovered as distillate. What are the flow rates of the distillate and bottoms product and the composition of the bottoms product?

2.3 100 kg of a mixture of sodium sulfate crystals ($Na_2SO_4 \cdot 10H_2O$) and sodium chloride (NaCl) is heated to drive away all the water. Final weight of the dry mixture is 58.075 kg. Calculate
(a) Mass of sodium sulfate crystal and sodium chloride in the original mixture
(b) Molar ratio of dry Na_2SO_4 and NaCl in the original mixture

2.4 A solution of NH_4Cl is saturated at 70°C. Calculate the temperature to which this solution must be cooled in order to crystallize out 45% of the NH_4Cl. The solubilities of NH_4Cl in water are

Temperature °C	70	10	0
Solubility g/100 g of water	60.2	33.3	29.4

2.5 Calculate the amount of H_2S in cubic meters measured at 49°C and at a pressure of 0.2 barG, which may be produced from 10 kg of FeS.

2.6 Determine the change in enthalpy in kJ when 64.40 kg of $Na_2SO_4 \cdot 10H_2O$ is dissolved in 108 kg of water at 25°C

Species	Heat of Formation at 25°C, kJ/kmol
Na_2SO_4 (infinite dilution)	-1.385×10^6
$Na_2SO_4 \cdot 10H_2O$ (infinite dilution)	-4.325×10^6
Water	-0.286×10^6
Molar enthalpy of Na_2SO_4 ($40H_2O$) =	-0.0109×10^6 kJ/kg mol at 25°C

2.7 A mixture of nitrogen and water vapor contains 25% by volume water vapor at a temperature of 80°C and 101.3 kPa pressure. The vapor pressure of water at 80°C = 47.3 kPa. Calculate (a) relative saturation and (b) % saturation of the mixture.

2.8 A solution of $NaNO_3$ at a temperature of 60°C contains 45% $NaNO_3$ by weight.
   (a) What is the percentage saturation of this solution? At 60°C, solubility of $NaNO_3$ = 124 g/100g of water.
   (b) Calculate the weight of $NaNO_3$ that will be crystallized if 100 g of this solution is cooled to 0°C.
   (c) Calculate the percentage yield of this crystallization process.
   (d) What is the molarity of the original solution? Specific gravity of solution = 1.3371 at 60°C.
   (e) What is the molality of the original solution?
   (f) What is the concentration of $NaNO_3$ in the original solution in grams per 100 g of water?

2.9 Air at 100 kPa pressure and a dry-bulb temperature of 93°C and wet-bulb temperature of 35°C is fed to a dryer. In the dryer 0.03 kmol of water evaporates per kmol of air fed to the dryer. If the vaporization of water in the dryer is adiabatic, calculate (a) the dry-bulb temperature, (b) the wet-bulb temperature, and (c) the percentage saturation of the air leaving the dryer.

2.10 1000 kg of a solution of 10% H_2SO_4 at 27°C is to be fortified to 50% strength by the addition of 98% H_2SO_4 which is at 21°C. How much heat is removed by the cooling system if the final temperature is to be 38°C? The enthalpies are as follows:

$$10\% \ H_2SO_4 \quad H = 12 \ \text{Btu/lb at } 27°C = 12 \times 2.326 = 27.912 \ \text{kJ/kg}$$
$$98\% \ H_2SO_4 \quad H = -3 \ \text{Btu/lb at } 21°C = -3 \times 2.326 = -6.978 \ \text{kJ/kg}$$
$$50\% \ H_2SO_4 \quad H = -84 \ \text{Btu/lb at } 38°C = -84 \times 2.326 = -195.384 \ \text{kJ/kg}$$

SOLUTIONS

2.1 Basic 100 kmol of fuel gas

(a) For an ideal gas, vol % and mol % are the same. Therefore, the molar composition of the gas is: Methane 85 mole %, Ethane 10.5%, and Nitrogen 4.5 mole %

(b) Composition in wt%

Component	mol %	kmols	mol. wt.	wt. in kg	wt%
Methane	85	85.00	16	1360	75.51
Ethane	10.5	10.50	30	315	17.49
Nitrogen	4.5	4.5	28	126	7.00
				1801	100.00

(c) Mol fractions $y_{methane} = 0.85$, $y_{ethane} = 0.105$, and $y_{nitrogen} = 0.045$
Average molecular weight of gas = $0.85 \times 16 + 0.105 \times 30 + 0.045 \times 28 = 18.01$

(d) Density at standard conditions
  (i) 1 kmol occupies 22.4 m^3 at std conditions of 0°C and 1 atm = 1.013 barA
  Therefore, density ρ = mol. wt./volume = $(18.01/22.4) = 0.804$ kg/m^3 at std conditions
  (ii) 1 kg = 2.2046 lb mass 1 m^3 = 35.3 ft^3
  Density in lb/ft^3, $\rho = (0.804 \times 2.2046)/(1 \times 35.3) = 0.0502$ lb_m/ft^3

(e) Total pressure = 1.013 barG = 1.013 + 1.013 = 2.026 barA = 2.026×100 = 202.6 kPa
Therefore, partial pressure of nitrogen = $0.045 \times 202.6 = 9.117$ kPa

2.2 Material Balance

Let F = feed, B = bottoms product, D = distillate, all in kg/h
Also y = mass fraction benzene in distillate = 0.98
x_B = mass fraction benzene in bottoms product
Basis: 2000 kg/h
Overall balance $F = B + D$
Benzene balance $Fx_f = yD + Bx_B$
Also from the given composition of distillate and the recovery of benzene in distillate

$$0.95\, Fx_f = 0.98\, D$$

By substitution of given values, the equations become

$$B + D = 2000 \tag{1}$$
$$(0.6\, F) = 0.98\, D + Bx_B \tag{2}$$
$$0.95(2000)(0.6) = 0.98\, D \tag{3}$$

From third equation, $D = \dfrac{0.95 \times 2000 \times 0.6}{0.98} = 1163.27$ kg/h

Substituting value of D in equation 1, $B = 836.73$ kg/h

Using values of B and D obtained above and substituting in equation 2

$$0.95\,(0.6F) = 0.98(1163.73) + 836.73 x_B$$

From which, $x_B = \dfrac{0.6(2000) - 0.98(1163.27)}{836.73} = 0.0717$ wt fraction

Therefore, bottoms composition is: benzene = 7.17 wt%, toluene = 92.83 wt%

Distillate = 1163.27 kg/h and bottoms product = 836.73 kg/h

2.3 Molecular weights: $Na_2SO_4 = 142$, $Na_2SO_4 \cdot 10H_2O = 322$, $H_2O = 18$, $NaCl = 58.5$

$$Na_2SO_4 \cdot 10H_2O \rightarrow Na_2SO_4 + 10H_2O$$
$$\quad\quad 322 \quad\quad\quad 142 \quad\quad 180$$

let x = kg of $Na_2SO_4 \cdot 10H_2O$ in the original mixture y = kg of NaCl in the original mixture

Material balance equations: (Basis of calculation: 100 kg of undried mixture)

1. Original mixture $\quad x + y = 100$
2. Dried mixture $\quad x - x\left(\dfrac{180}{322}\right) + y = 58.075$

Subtracting Equation 2 from Equation 1 gives

$$\left(\dfrac{180}{322}\right) x = 41.925 \text{ or } x = 41.925 \times (322/180) = 75 \text{ kg}$$
$$y = 100 - 75 = 25 \text{ kg}$$

Dry Na_2SO_4 in mixture = $(142/322) \times 75 = 33.075$ kg = $33.075/142 = 0.2329$ kg mol

NaCl in the mixture = $25/58.5 = 0.42735$ kg mol

Molar ratio of dry Na_2SO_4 to NaCl = $0.2329/0.42735 = 0.545$

2.4 At 70 °C, solubility of NH_4Cl = 60.2 g/100 g of water
45% of NH_4Cl to be crystallized.
NH_4Cl in solution = $(1 - 0.45) \times 60.2 = 33.1$ g
Solubilities of NH_4Cl at 0 and 10 °C are 29.4 and 33.3 g/100 g of water respectively. By interpolation, the temperature at which the solubility is 33.1 g/100 g of water is 9.5 °C.

2.5 $\text{FeS} \times H_2S$
88×34
Amount of H_2S produced = $(34/88) \times 10 = 3.864$ kg per 10 kg of FeS
Since no other data are given, assume ideal gas behavior.
Pressure = $0.2 + 1.013 = 1.213$ barA, $T = 50 + 273 = 323$ K

kg mol of H_2S = 3.864/34 = 0.113647 kg mol
Volume at std conditions = 0.113647(22.4) = 2.546 m^3
Volume at 50°C and 1.213 bar abs, V =

$$(2.546)\left(\frac{323}{273}\right)\left(\frac{1.013}{1.213}\right) = 2.516 \text{ m}^3.$$

2.6 64.4 kg of $Na_2SO_4 \cdot 10H_2O$ = 64.4/322 = 0.2 kg mol
108 kg of water = 108/18 = 6 kg mol
Basis of calculation = 1 kg mol of $Na_2SO_4 \cdot 10H_2O$
Corresponding amount of water is (1/0.2) × 6 = 30 kg mol of water
Heat of reaction for the reaction $Na_2SO_4 + 10H_2O \rightarrow Na_2SO_4 \cdot 10H_2O$ –
1.385 × 10^6 10(– 0.286 × 10^6)   – 4.325 × 10^6
ΔH_R = –4.325 × 10^6 – (–1.38 × 10^6) – (10 × 0.286 × 10^6) = –0.085 × 10^6 kJ/kg mol
This is molal enthalpy of $Na_2SO_4 \cdot 10H_2O$ relative to Na_2SO_4 and H_2O at 25°C.
Molal enthalpy of $Na_2SO_4 \cdot H_2O$ relative to Na_2SO_4 and H_2O is =
– 0.0109 × 10^6 kJ/kg mol at 25°C. Then for the reaction
$Na_2SO_4 \cdot 10H_2O + 30H_2O \rightarrow Na_2SO_4 \cdot 40H_2O$
ΔH = –0.0109 × 10^6 – (–0.085 × 10^6) = + 0.0741 × 10^6 kJ/kg mol
Therefore, change in enthalpy when 0.2 kg mol of $Na_2SO_4 \cdot 10H_2O$ is dissolved in 6 kg mol of water ΔH = (0.2) × 0.0741 × 10^6 = 14820 kJ

2.7 (a) Partial pressure of water vapor = 0.25 (101.3) = 25.33 kPa
Vapor pressure of water = 47.3 kPa
Relative saturation = (p_w/p) × 100 = (25.33/47.3) × 100 = 53.55%

(b) Basis 1 kg mol of mixture
Moles of nitrogen in mixture = 1.0 – 0.25 = 0.75 kg mol
Moles of vapor in the mixture = 0.25 kg mol
Moles of vapor/mole of vapor-free gas = 0.25/0.75 = 0.333 kg mol
Vapor mole fraction at saturation (80°C) = 47.3/101.3 = 0.467
Moles of vapor/Mole of vapor-free gas at saturation = 0.467/(1 – 0.467) = 0.8762
Therefore, % saturation of mixture = (0.333)/(0.8762) × 100 = 38%

2.8 (a) At 60°C, solubility of $NaNO_3$ = 124g/100g of water
Initial solution is 45% $NaNO_3$ by weight, or (45/55) = 81.82g/100g of water
Therefore, % saturation = 81.82/124 = 66%

(b) Solubility of $NaNO_3$ at 0°C = 73 g per 100 g of water or $NaNO_3$ is 42.2% by weight.
Original solution has 50 g $NaNO_3$
Let x be the amount of $NaNO_3$ precipitated on cooling
By material balance on $NaNO_3$,
100 × (0.45) = x + 0.422 (100 – x)
Which on solving gives x = 4.84 g

(c) % yield = (4.84/45) × 100 = 10.76%

(d) MW of $NaNO_3$ = 85
$NaNO_3$ in solution = 45/85 = 0.53 gmol in 100 g of solution
Specific gravity of 45% $NaNO_3$ solution = 1.3371 at 60°C
Volume of solution = 100/1.3371 = 74.79 cc
Therefore, molarity = (0.53/74.79) × 1000 = 7.09 gmol/liter of solution

(e) Molality = (0.53/(100 − 45) × 1000 = 9.64 gmol/1000 g of water

(f) Original solution contains 45 g of $NaNO_3$ in 55 g of water
Concentration of $NaNO_3$ = (45/55) × 100 = 81.82g/100g of water

2.9 At 93°C dry-bulb and 35°C wet-bulb temperature, molal humidity is 0.02 kg mol/kg mol dry air. This is read from psychrometric chart given in *Chemical Engineering License Review*[1].

Water evaporated in dryer = 0.03 kg mol/kg mol dry air

Therefore, humidity of air leaving the dryer = 0.02 + 0.03 = 0.05 kg mol/kg mol dry air

Since vaporization is adiabatic, wet-bulb temperature remains the same and equals 35°C

Dry-bulb temperature at molal humidity of 0.05 kg mol/kg mol dry air and wet-bulb temperature of 35°C is read from the same chart and is equal to 47.5°C

Saturation humidity at 47.5°C dry-bulb = 0.12 kg mol/kg mol dry air

Therefore, % saturation = (0.05/0.12) × 100 = 41.7%

2.10 For the solution of this problem, use the enthalpy composition diagram[2,3] for sulfuric acid and water system.

Material Balance

Let x be the amount of 98% acid added. Then balance on H_2SO_4 gives

$$0.98x + 0.1(1000) = 0.5(1000 + x)$$

Therefore,

$$x = 833.3 \text{ kg of } 98\% \text{ } H_2SO_4$$

Energy balance gives the heat to be removed by the cooling system.

$$Q = -195.384(1000 + 833.3) - 1000(27.912) - 833.3(-6.978)$$
$$= -380295 \text{ kJ}$$

Heat to be removed by the cooling system = 380295 kJ

CHAPTER 3

Thermodynamics

OUTLINE

TERMINOLOGY 30

THERMODYNAMIC PROPERTIES 30
Mass Balance ■ Energy Balance (First Law of Thermodynamics)

SECOND LAW OF THERMODYNAMICS 32
Entropy ■ Carnot Principle ■ General Statement of Second Law of Thermodynamics ■ Calculation of Entropy Changes

PROPERTIES OF IDEAL GASES 35
Thermodynamic Relations for an Ideal Gas

PROPERTIES OF REAL GASES 36
Fugacity and Fugacity Coefficient ■ Reduced Conditions ■ Property Charts ■ Clapeyron Equation

THIRD LAW OF THERMODYNAMICS 42

CYCLIC PROCESSES 42
The Carnot Cycle ■ Refrigeration ■ Rankine Cycle

HEATS OF MIXING AND ENTHALPY OF A SOLUTION 45

GIBBS' PHASE RULE 46

REFERENCES 46

Thermodynamics deals with the transformation of energy from one form to another in macro systems. Experience has shown that all energy transformations occur under certain restrictions, which are known as laws of thermodynamics. These laws together with definitions of the properties of materials allow a chemical engineer to study a wide variety of problems connected with physical and chemical processes. The most important problems are the estimation of heat and work requirements of a process, determination of chemical reaction equilibrium, and equilibrium for the transfer of chemical species between phases. Because of the importance of thermodynamics in the analysis of physical and chemical processes, and the calculation of properties of pure materials, FE examinees should expect at least six problems in this area.

TERMINOLOGY

We begin this review of thermodynamics with some basic definitions.

System—a portion of the universe (e.g., a substance or a group of substances) set apart for study.

Closed system—a system with constant mass. There is no exchange of matter between the system and the surroundings. A closed system is thermally isolated when its enclosing walls allow no flow of heat into or out of the system. It is mechanically isolated if it is enclosed by rigid walls. It is completely isolated if neither matter nor energy in any form can be added to or removed from the system.

Open system—a system with variable mass (mass is transferred across the boundaries of the system).

Reversible process—one that begins from an equilibrium state of a system and proceeds under conditions of balanced forces in such a manner that its direction can be reversed by applying an infinitesimal driving force in the opposite direction, restoring both the system and its surroundings to their initial equilibrium state. In a reversible process, there is no degradation of energy taking place and availability of the energy of the combined system and its surroundings is the same before or after the process as the system and its surroundings are restored to their original equilibrium state.

As an example of a reversible process, one may consider the vaporization of a liquid (enclosed in a cylinder with a frictionless piston and in contact with an isothermal reservoir) under its own vapor pressure. At any time, an infinitely small increase in pressure on the piston will cause condensation, whereas an infinitely small decrease in pressure will result in vaporization of the liquid. This is, however, an ideal situation, as the concepts of a frictionless piston and an isothermal reservoir are assumptions only and are far removed from actual situations.

Actual processes are invariably irreversible and therefore there is in reality a decrease in available energy. A few examples of irreversible processes are (a) flow of heat from one body to the other under the influence of a finite temperature difference, (b) free expansion of a gas, (c) mixing of hot and cold fluids, and (d) a spontaneous chemical reaction.

In a later section, reversibility will again be reviewed when the second law of thermodynamics and the concept of entropy are postulated.

THERMODYNAMIC PROPERTIES

Any measurable characteristic of a system in terms of fundamental dimensions is called its property, for example, temperature, pressure, mass, area, volume, surface tension, and so on.

Intensive properties, such as pressure and temperature, are independent of mass.

Extensive properties depend on the mass of the system. Volume, internal energy, and enthalpy are examples of extensive properties. However, when these extensive properties are based on a unit such as a unit mass or mol, they become intensive properties because their values do not depend upon the total material actually present in the system. Specific volume or density, internal energy per unit mass or mol, and enthalpy per unit basis are also examples of intensive properties.

State properties depend upon the thermodynamic state of the system and are independent of the path the system takes to reach that state. Some state properties are: pressure P, temperature T, specific volume V, internal energy U, enthalpy H, entropy S, Gibbs free energy G, Helmholtz free energy A, and heat capacities at constant pressure C_P and at constant volume C_V.

Mass and energy transfers are closely related. Mass balance (principle of conservation of mass) and energy balance (principle of conservation of energy or first law of thermodynamics) equations besides the second law of thermodynamics are the foundations of any thermodynamic analysis of processes. Mass and energy balance equations are reviewed in Chapter 2. However, because of their importance in thermodynamic analysis, we briefly review them again in this chapter along with the second law of thermodynamics and some related concepts and equations. We will also look briefly at the third law of thermodynamics and its usefulness.

Mass Balance

Application of the accounting principle to mass of a system over a chosen accounting period yields the mass balance equation:

$$M_E - M_B = \sum M_I - \sum M_O + \sum M_P - \sum M_C$$

The mass balance equation has already been discussed in detail in Chapter 2. In the analysis of ordinary physical and chemical processes, the $\sum M_C$ term is equal to zero and therefore we will be dealing with the more common form of the equation

$$M_E - M_B = \sum M_I - \sum M_O + \sum M_P$$

Energy Balance (First Law of Thermodynamics)

Energy balance is based on the principle of conservation of energy, or the first law of thermodynamics.

Application of the accounting principle to energy of a system over a chosen accounting period yields:

$$(U + PE + KE)_E - (U + PE + KE)_B = \sum(H + PE + KE)_I \\ - \sum(H + PE + KE)_o + Q + W - \sum E_C$$

where
    U = internal energy
    PE = potential energy
    KE = kinetic energy
    H = enthalpy
    Q = heat transferred across the boundary of the system during the accounting period
    W = all work that crosses the system boundaries during the accounting period
    $\sum E_C$ = energy conversions due to atomic transmutations

In ordinary non-nuclear reactions or for ordinary physical and chemical processes $\sum E_C = 0$. In this case, the energy balance reduces to a more practical form:

$$(U + PE + KE)_E - (U + PE + KE)_B = \sum(H + PE + KE)_I - \sum(H + PE + KE)_o + Q + W$$

Terms in the energy balance equation are defined in Chapter 2.

When Q enters the system, it is taken as positive. When it leaves the system it is taken as negative. Work done on the system is positive, whereas work done by the system is negative.

$$\text{Work} = \text{force} \times \text{distance}$$

When fluid pressure is the only force acting on the system, $W = \int_{V_1}^{V_2} PdV$.

In many applications, the work term will consist of W_s, shaft work, which is the work done on or by the fluid flowing through a system and transmitted by a shaft. This is the work that is transferred between the system and the surroundings through the shaft. In addition to this work, there is work exchanged between the fluid mass chosen as a system and the fluid on either side of it. For a steady state flow, the energy equation simplifies to the following form:

$$(H + PE + KE)_O - (H + PE + KE)_I = Q + W_S$$

or

$$\Delta(H + PE + KE) = Q + W_S$$

For many applications, the potential and kinetic energy terms are negligible compared to others and can be neglected. In such cases, the energy balance reduces simply to

$$\Delta H = Q + W_S$$

The mass and energy balance equations permit to relate the mass, work, and heat flows of a system to the change in its thermodynamic state. However, these two balance equations are not adequate to solve all energy flow problems of interest. Natural processes are irreversible. To account for the irreversibility of natural processes, the second law of thermodynamics has been postulated and an additional thermodynamic state variable called *entropy* has been defined. These are reviewed in the next sections.

SECOND LAW OF THERMODYNAMICS

The concept of reversibility of a process is the basis on which the second law of thermodynamics is postulated. There are various statements of this law. One statement is that *all the spontaneous processes are invariably irreversible to some extent and are accompanied by a degradation of energy.*

Heat cannot be converted into work quantitatively and cannot be transferred from a lower temperature to a higher temperature without the aid of an external agency. This principle is covered by another statement of the second law of thermodynamics: *No practical engine can convert heat into work quantitatively nor it is impossible for a self-acting machine unaided by an external agency to transfer heat from a lower temperature to a higher one.*

The validity of the second law rests completely on experimental evidence.

Entropy

Mass and energy balances are equalities, which are not sufficient to deal with all the thermodynamic energy flow problems. Specifically, these are not adequate to determine heat and work flows for a system undergoing a change of state or to take into account the unidirectional character of natural processes and resulting inequalities. A new thermodynamic variable or state property called *entropy* (S) is therefore defined that allows to write another balance equation to account for the limitations imposed by the second law of thermodynamics.

Entropy is defined as a state property, which changes due to the heat flow but not the work flow crossing the system boundaries. The rate of entropy change is expressed as

$$dS = \frac{d'Q}{T}$$

where
 S = entropy
 $d'Q$ = incremental heat flow that crosses the system boundaries
 T = absolute temperature of the system

Like internal energy U or enthalpy H, the entropy is an intensive state property.

Carnot Principle

The following two statements together constitute the Carnot principle:

1. *The efficiencies of two reversible heat engines operating between the same temperature levels are equal.*

2. *The efficiency of an irreversible heat engine is always less than that of a reversible engine.*

Although heat and work are both forms of energy, there is a difference between them. This difference lies in the fact that heat can be converted into useful work by an engine only partially. On the other hand, work can be spontaneously converted to heat completely at any temperature. One deduction from the second law of thermodynamics is that any engine that performs work by absorbing heat at a higher temperature and rejecting some of it at a lower temperature will convert the maximum amount of heat energy into useful work if it acts reversibly. Consequently, any engine that acts irreversibly converts a smaller fraction of heat into useful work and therefore is less efficient compared to a reversible engine.

One aspect of the Carnot principle is that the efficiency of a reversible engine depends only on temperature levels at which heat is absorbed and rejected and is independent of the medium used. Thus

$$\text{Efficiency, } \eta_{rev} = \frac{W_{rev}}{Q_1} = \frac{Q_1 - Q_2}{Q_1} = \frac{T_1 - T_2}{T_1}$$

For an engine undergoing a reversible Carnot cycle, $\oint \frac{dQ_R}{T} = 0$. This is the change in entropy and it is a state property. The Carnot cycle is discussed later in this chapter.

Work done by the engine using a finite temperature difference is

$$W = Q_1 \left(\frac{T_1 - T_2}{T_1} \right) - T_2 S_P$$

where
 S_P = entropy produced
 S_p = 0 for a reversible process

$S_p > 0$ for an irreversible process

General Statement of Second Law of Thermodynamics

Once the entropy, an intensive state property, is defined a general statement of the second law of thermodynamics in mathematical terms can be given as follows:

$$S_E - S_B = \Sigma S_t + \Sigma \frac{Q}{T_B} + S_P - S_C \qquad S_P \geq 0$$

where
S_t = entropy changes due to transfer of mass across the boundary of the system
S_C = entropy changes occurring in a system due to atomic transmutations or relativistic effects

Note that the entropy production term is equal to zero for a reversible process and positive for an irreversible process. In physical and chemical processes, there are no atomic transmutations and relativistic effects, and therefore, $S_C = 0$ for these processes. Hence the common more practical form of the entropy balance equation is

$$S_E - S_B = \Sigma S_t + \Sigma \frac{Q}{T_B} + S_P \qquad S_P \geq 0$$

Calculation of Entropy Changes

Quantitative application of the second law of thermodynamics requires the calculation of entropy changes. Relationships between various thermodynamic functions can be established using the first and second laws of thermodynamics. Some of these relations are given later.

For a constant mass and composition system undergoing a reversible process with pressure as the only external force

$$dU = TdS - PdV$$

Also $$dH = TdS + VdP$$

Helmholtz function $A = U - TS$ and $dA = -PdV - SdT$

Gibb's free energy $G = H - TS$ and $dG = VdP - SdT$

These equations provide the basis for calculating entropy changes for real fluids.

The following examples will demonstrate the procedures used to calculate the entropy changes.

Energy absorption at constant temperature (no chemical reaction): $\Delta S = \dfrac{Q}{T}$

Isothermal phase change: $\Delta S = \dfrac{\lambda}{T}$

Heating or cooling without chemical change: $\Delta S = \displaystyle\int_{T_1}^{T_2} \dfrac{d'q}{T}$

For a constant-pressure condition, when heat capacity is independent of temperature

$$\Delta S = \int_{T_1}^{T_2} \frac{nC_p dT}{T} = nC_p \ln \frac{T_2}{T_1}$$

If the heat capacity varies with the temperature, the equation relating heat capacity with temperature is used and the equation is then integrated.

Availability function $B = -[\Delta H - T_o \Delta S]_{T_1}^{T_2}$

For a flow system, $\Delta B = [\Delta H - T_0 \Delta S]_{T_1}^{T_2}$

When *KE* and *PE* changes occur, for a flow system,

$$\Delta B = \left[\Delta H - T_0 \Delta S + \frac{\Delta u^2}{2gc} + \frac{g}{gc} \Delta z \right]_{T_1}^{T_2}$$

PROPERTIES OF IDEAL GASES

Properties of an ideal gas are defined by the requirement that the *PVT* relationship is given by the equation

$$PV = nRT$$

where
- P = pressure
- T = temperature
- V = volume
- n = number of moles
- R = gas law constant

Boyle's law states that $P_1 V_1 = P_2 V_2$ at constant T and n.

Charles' law states that $\dfrac{V_1}{T_1} = \dfrac{V_2}{T_2}$ at constant P and n.

Dalton's law of partial pressures and *Amagat's law* of partial volumes also apply to ideal gases.

Thermodynamic Relations for an Ideal Gas

Constant volume (isometric) process: $d\hat{U} = dQ = C_v dT$ or $\Delta \hat{U} = Q = \int C_v dT$

Constant pressure (isobaric) process: $d\hat{H} = dQ = C_p dT$ or $\Delta \hat{H} = Q = \int C_p d$

Constant temperature (isothermal) process: $Q = W = RT \ln \dfrac{\hat{V}_2}{\hat{V}_1} = RT \ln \dfrac{P_1}{P_2}$

Adiabatic ($dQ = 0$) process: $\dfrac{T_2}{T_1} = \left(\dfrac{\hat{V}_1}{\hat{V}_2} \right)^{k-1} = \left(\dfrac{P_2}{P_1} \right)^{(k-1)/k}$ or $P\hat{V}^k$ = constant

where
k = ratio of specific heats = $\dfrac{C_p}{C_v}$

Work for adiabatic process: $W = \dfrac{RT_1}{k-1} \left[1 - \left(\dfrac{P_2}{P_1} \right)^{(k-1)/k} \right]$

Polytropic compression of an ideal gas: $\dfrac{n-1}{n} = \dfrac{k-1}{\eta_p \cdot k}$ η_P = polytropic efficiency

Then

$$W_P = \frac{RT_1}{n-1}\left[1 - \left(\frac{P_2}{P_1}\right)^{(n-1)/n}\right]$$

Also

$$\frac{T_2}{T_1} = \left(\frac{P_2}{P_1}\right)^{(n-1)/n} = \left(\frac{\hat{V}_1}{\hat{V}_2}\right)^{n-1}$$

Entropy changes for an ideal gas: when an ideal gas is compressed from P_1, T_1, V_1 to P_2, T_2, V_2, the entropy change is given by

$$\Delta S_T = \Delta S_1 + \Delta S_2 = \int_{T_1}^{T_2} \frac{C_P}{T} dT - R \ln \frac{P_2}{P_1}$$

(Change temperature at constant pressure and then change the pressure reversibly and isothermally.)

By changing temperature at constant volume and then changing volume reversibly and isothermally,

$$\Delta S_T = \int_{T_1}^{T_2} \frac{C_V}{T} dT + R \ln \frac{\hat{V}_2}{\hat{V}_1}$$

PROPERTIES OF REAL GASES

For real gases, the ideal gas law does not hold. Therefore other equations of state are proposed. Some of these equations are the following:

- Van der Waal's equation:

$$\left(P + \frac{a}{\hat{V}^2}\right)(\hat{V} - b) = RT$$

where
  a and b are Van der Waal's constants
In terms of critical constant

$$a = \frac{27R^2T_c^2}{64P_c} = (9/8)RT_c\hat{V}_c \quad \text{and} \quad b = \frac{RT_c}{8P_c} = \frac{\hat{V}_c}{3}$$

where
  b = excluded volume/mole

- Redlich-Kwong equation:

$$P = \frac{RT}{\hat{V}-b} - \frac{a}{T^{1/2}\hat{V}(\hat{V}+b)}$$

where $\quad a = \dfrac{0.4278 R^2 T_c^2}{P_c} \quad$ and $\quad b = 0.0867 \dfrac{RT_c}{P_c}$

- Van der Waal's equation in terms of reduced conditions:

$$\left(P_r + \frac{3}{\hat{V}_r}\right)\left(\hat{V}_r - \frac{1}{3}\right) = \frac{8}{3}T_r$$

- Beattie-Bridgeman Equation of state:

$$Pv^2 = RT\left[v + B_0\left(1 - \frac{b}{v}\right)\right]\left(1 - \frac{c}{vT^3}\right) - A_0\left(1 - \frac{a}{v}\right)$$

- The constants in each of the preceding equations are empirically determined.

Fugacity and Fugacity Coefficient

Fugacity of a pure substance is defined by the equation

$$(dG)_T = RT\, d\ln f_i \quad \text{at constant temperature}$$

For ideal gas, this equation becomes

$$(dG)_T = \hat{V}dP = RT\frac{dP}{P} = RTd\ln P$$

A further restriction on the definition of f_i^V is $\lim\limits_{P\to 0} \dfrac{f_i^V}{P} = 1$.

The ratio of the fugacity to its pressure or $\dfrac{\hat{f}_i^L}{P}$ is called the *fugacity coefficient* ϕ_i^V of the component i in the vapor phase.

Likewise the fugacity $\dfrac{\hat{f}_i^L}{P}$ of a component in solution is defined by the equations

$$d\bar{G}_i = RTd\ln \hat{f}_i^L \quad \text{at constant temperature}$$

and

$$\lim_{P\to 0} \frac{\hat{f}_i^L}{x_i P} = 1$$

For a component in solution, the *fugacity coefficient* is defined in the same manner as that for vapor phase and is given by

$$\hat{\phi}_i^L = \frac{\hat{f}_i^L}{x_i P}$$

The fugacity f of a solution as a whole is also defined in the same manner as the fugacity of a pure material.

$$dG = RT \, d \ln f \quad \text{at constant temperature}$$

and

$$\lim_{P \to 0} \frac{f}{P} = 1$$

In this case G is the molar Gibbs free energy of the solution.

By combining the equation for fugacity with that for ideal gas and noting that at constant temperature $\left(\dfrac{d \ln f_i^v}{d \ln P}\right) = z$ (compressibility), one can derive the following relation:

$$d \ln \frac{f_i^V}{P} = [(z-1) d \ln P] \quad \text{at constant } T$$

At $P = 0$, fugacity and pressure are equal by definition. Using this fact and writing the previous equation in terms of reduced conditions and by integration from $P = 0$ to P_r, we get

$$\ln \frac{f_i^V}{P} = \left[\int_0^{P_r} \frac{(z-1) dP_r}{P_r}\right]$$

Fugacity can be determined using PVT data or an equation of state. If the law of corresponding states is used

$$dG = RT d \ln f_i^V = \frac{zRT}{P} dP \quad \text{at constant } T$$

$$\text{or } \ln f_i^v = \int_0^P z \frac{dP}{P}$$

The integral on the right-hand side of the equation can be evaluated by numerical integration using generalized z charts. Hougen et al. have given a table of fugacity coefficients, ϕ_i versus P_r with T_r as parameter for substances having critical compressibility $z_c = 0.27$. A table of data and a correction equation allow the computation of the fugacity coefficients for other materials having different critical compressibility factors.

Fugacities in ideal solutions and standard states

The fugacity coefficients of a component in a solution and its pure state are related by

$$\ln \frac{\hat{\phi}_i}{\phi_i} = \frac{1}{RT} \int_0^P (\hat{V}_i - V_i) dP$$

If $\hat{\phi}_i$ and ϕ_i are replaced by their definitions, one obtains

$$\ln \frac{\hat{f}_i}{f_i x_i} = \frac{1}{RT} \int_0^P (\hat{V}_i - V_i) dP$$

For an ideal solution, $\hat{V}_i = V_i$, therefore $\ln \dfrac{\hat{f}_i^{id}}{f_i x_i} = 0$ and then $\hat{f}_i^{id} = f_i x_i$.

This expression shows that for an ideal solution, the fugacity of a component in solution is equal to its mol fraction times its fugacity in pure state. This is known as the *Lewis-Randall rule*.

A more general definition of an ideal solution is based on the concept of *standard states*. Based upon a standard state, the fugacity of component i in an ideal solution is defined by

$$\widehat{f}_i^{id} = f_i x_i.$$

where f_i^o is the fugacity of component i in a standard state at the same temperature and pressure as that of the mixture. The standard state can be either the actual state of pure i or it could be a hypothetical state. If the actual state of pure i is chosen, $f_i^o = f_i$. Two convenient states, one based on the Lewis-Randall rule and the other based on Henry's law are defined as follows:

Based on the Lewis-Randall rule

$$\lim_{x_i \to 1} \frac{\widehat{f}_i}{x_i} = f_i \quad \text{here standard state fugacity is } f_i$$

Based on Henry's law

$$\lim_{x_i \to 0} \frac{\widehat{f}_i}{x_i} = f_i \quad \text{here standard state fugacity is } k_i$$

With standard states defined in the previous manner, the fugacity of ideal solution will be given by

$$\widehat{f}_i^{id} = x_i f_i \quad \text{(based on Lewis-Randall rule)}$$

where, f_i is the fugacity of component i in the standard state.

The other standard state results in the following expression for the fugacity of a component in solution

$$\widehat{f}_i^{id} = x_i k_i \quad \text{(based on Henry's law)}$$

This relation is valid for values of x_i near zero. In this case, the standard state fugacity is k_i and $k_i = f_i$.

Fugacity and phase equilibria

In a multicomponent system at equilibrium, the fugacity of each component is the same in all phases. Thus in a two-phase system of a multicomponent liquid mixture in equilibrium with its vapor

$$\widehat{f}_i^V(T,P,y) = \widehat{f}_i^L(T,P,x)$$

Fugacity of a component i in a liquid mixture is usually calculated by using the relation

$$\widehat{f}_i^L = x_i \gamma_i f_i^L$$

where

f_i^L = fugacity of pure liquid component i
γ_i = activity coefficient of component i
x_i = mol fraction of component i in liquid mixture

The fugacity of the pure component can be calculated as

$$f_i^L = \phi_i^{sat} P_i^{sat} \exp\{v_i^L(P - P_i^{sat})/RT\}$$

where

ϕ_i^{sat} = fugacity coefficient of pure saturated i
P_i^{sat} = saturation pressure of pure component i
v_i^L = molar specific volume of pure liquid component i

Very often the system pressure is low and close to atmospheric. In this situation

$$f_i^L = P_i^{sat}$$

The calculation of activity coefficient γ_i is reviewed in Chapter 14, "Distillation."

For the case of vapor in equilibrium with the liquid mixture, the fugacity can be calculated by the relation

$$\hat{f}_i^V = y_i \hat{\phi}_i P$$

where

y_i = mol fraction of component i in vapor
$\hat{\phi}_i$ = fugacity coefficient of component i in vapor
P = system pressure

The fugacity coefficient $\hat{\phi}_i$ can be calculated by the equation as

$$\ln \hat{\phi}_i = \int_0^P \left(\frac{\bar{v}_i}{RT} - \frac{1}{P} \right) dP$$

If *PVT* data are available, the integral on the right-hand side can be evaluated and, using this value of $\hat{\phi}_i$, $\hat{f}_i^V$, can be calculated by the equation $\hat{f}_i^V = y_i \hat{\phi}_i P$.

Reduced Conditions

Reduced conditions of temperature, pressure, and volume are defined by the following relations.

$$T_r = T/T_c, \quad P_r = P/P_c, \quad \text{and } V_r = V/V_c$$

If pure substances exist at the same reduced conditions of temperature and pressure, they are said to be in corresponding state. Pure gases in corresponding states have the same compressibility factors at the same reduced conditions of temperature and pressure. According to this principle, the deviations of thermodynamic properties of different pure fluids would show the same departure from the properties of these fluids in their ideal gaseous state at the same reduced conditions of temperature and pressure. Based on this principle, generalized charts have been prepared for compressibility factors for gases and properties such as the fugacity derived therefrom and for departures of enthalpy, entropy, and heat capacity from ideal gas behavior. These generalized charts were later improved upon by adding a new parameter z_c, the critical compressibility factor which is reviewed next.

Compressibility factor

The equation of state is written as $PV = znRT$ where z is the compressibility factor. It is a function of pressure, temperature, and the nature of the gas. The *theorem of corresponding states* postulates that all pure gases have the same compressibility factors when measured at the same reduced conditions of temperature and pressure. Generalized z charts have been developed for the compressibility factors.

$$\text{Critical compressibility factor} = z_c = \frac{P_c V_c}{RT_c}$$

$$\text{Virial equation of state: } z = \frac{P\hat{V}}{RT} = 1 + \frac{B}{\hat{V}} + \frac{C}{\hat{V}^2} + \cdots$$

$$\text{Other form of virial equation is: } z = \frac{P\hat{V}}{RT} = 1 + B'P + C'P^2 + D'P^3 + \ldots$$

For ideal gas, $dG_T = \hat{V}dP = RT\dfrac{dP}{P} = RTd\ln P$ at constant temperature

For non-ideal substances, fugacity is defined by $dG_{Ti} = RTd\ln f_i$ and $\lim_{p \to 0}(f_i/P) = 1.0$

For an ideal gas, this becomes $dG_T = \hat{V}\, dP = RT\dfrac{dP}{P} = RTd\ln P$ at constant temperature

Property Charts

Commonly used property charts include the following:

- Pressure-volume (*PV*) chart with *T* as parameter
- Temperature-entropy (*T-S*) diagram
- Pressure-enthalpy (*P-H*) diagram
- Enthalpy-entropy (*H-S*), also called Mollier diagram

The Mollier diagram is more useful in solving the turbine and compressor problems. The *P-H* diagram is useful in solving refrigeration problems, whereas the *T-S* diagram is useful in analysis of engines.

Clapeyron Equation

Phase changes for a pure substance occur at constant temperature and pressure. Melting, freezing, vaporization, and sublimation are examples of phase transition. Molar or specific properties, such as specific volume, internal energy, enthalpy, and entropy, change abruptly due to a phase change. Gibbs free energy function *G* is an exception. Generally, for two coexisting phases 1 and 2 of a pure substance, $G^1 = G^2$. Using the definition of *G* and the changes in entropy and volume that occur due to the phase change, the following Clapeyron equation applicable to all phase transitions can be derived as

$$\frac{dP^{sat}}{dT} = \frac{\Delta H^{1,2}}{T\Delta V^{1,2}}$$

where
- $\Delta H^{1,2}$ = enthalpy change of phase transition
- $\Delta V^{1,2}$ = volume change of phase transition

For the case of vaporization from a pure liquid, the Clapeyron equation is written as

$$\frac{dP}{dT} = \frac{\Delta H_V}{T\Delta V}$$

where
- ΔH_V = heat of vaporization
- V = volume change because of phase change

THIRD LAW OF THERMODYNAMICS

We conclude this review of mass, energy, and entropy balance concepts for thermodynamics with the third law of thermodynamics. Nerst and Plank postulated that at absolute zero temperature, the absolute entropy of a purely crystalline substance is zero. The absolute entropy of a pure substance at constant pressure can be calculated from

$$\hat{S} = \int_0^T \frac{C_{ps}}{T} dT + \frac{\Delta H_f}{T_f} + \int_{T_l}^{T_b} \frac{C_{pl}}{T} dT + \frac{\Delta \hat{H}v}{T_b} + \int_{T_b}^{T} \frac{C_{pg}}{T} dT$$

In order to calculate the absolute entropy of a pure substance, it is necessary to have data on the heats of phase transition and the specific heats of solid, liquid, and vapor as a function of temperature.

CYCLIC PROCESSES

A cyclic process consists of a series of operations on a working fluid periodically repeated in the same order. Operations may be repeated on the same mass of a substance, which is recirculated, or a new but equal mass of the same substance for each cycle. The objective is to convert heat into useful work efficiently. A coal-fired power plant, for example, uses steam as working fluid. Internal combustion engines, such as Otto engine or combustion gas turbine, use direct evolution of heat within the work-producing device.

A cyclic process that is completely reversible must necessarily consist of individual steps, each of which is reversible. To obtain maximum work from an engine operating between two temperature levels T_1 and T_2, all the steps must be carried out reversibly so that entropy generation is equal to zero.

The Carnot Cycle

The Carnot engine is a hypothetical idealized device that carries out a series of reversible operations on an ideal gas as working fluid enclosed in a cylinder equipped with a frictionless piston. Since these steps are reversible, it gives the maximum possible conversion of heat into work in a cyclic process. Hence it is used as a standard in evaluating the efficiency and performance of actual engines. The Carnot cycle is depicted on PV and TS diagrams in Figure 3.1.

Network is given by

$$W_{rev} = \Sigma \oint P dV = Q_1 - Q_2 \quad \Delta S_T = \Delta S_1 + \Delta S_2 = 0$$

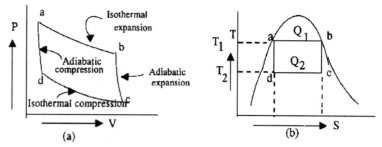

Figure 3.1 Carnot cycle on (a) PV diagram and (b) TS diagram.

Therefore, the thermodynamic ideal efficiency of the Carnot cycle is

$$\frac{Q_1 - Q_2}{Q_1} = \frac{T_1 - T_2}{T_1} = \frac{W_{net}}{Q_1}$$

Refrigeration

Refrigeration is a process in which a working fluid is used to absorb heat from a low temperature source and to reject it at a higher temperature. A ton of refrigeration is the heat removal rate of 200 Btu/min, 12,000 Btu/h, or 288,000 Btu/d. One ton of refrigeration is also equal to the latent heat of melting of one ton of ice in 24 hours.

The coefficient of performance, β, for refrigeration is defined as:

$$\beta = \frac{\text{heat removed from low temperature region}}{\text{work of compression}}$$

Reversed Carnot Cycle

The ideal refrigeration cycle is a reversed Carnot cycle. It is shown in Figure 3.2.

The coefficient of performance, β of the ideal refrigeration cycle is given by

$$\beta = \frac{\text{heat absorbed in evaporator}}{\text{work done on the fluid}} = \frac{Q_1}{-W} = \frac{Q_1}{Q_2 - Q_1} \frac{T_1 \Delta S}{(T_2 - T_1)\Delta S} = \frac{T_1}{T_2 - T_1}$$

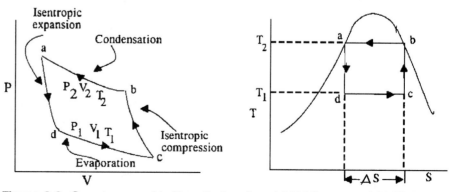

Figure 3.2 Carnot reversed (refrigeration) cycle on (a) PV diagram and (b) TS diagram.

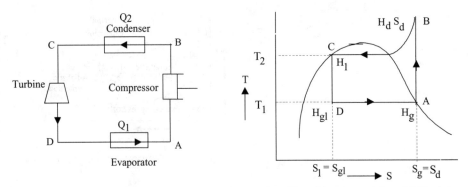

Figure 3.3 Vapor compression refrigeration cycle with turbine expansion. (a) Sketch of a system (b) TS diagram.

Vapor compression with turbine expansion cycle

Vapor compression with turbine expansion cycle is shown in Figure 3.3.

The coefficient of performance β of the vapor compression refrigeration cycle with turbine expansion is given by the following relation:

$$\text{Coefficient of performance, } \beta = \frac{H_g - H_{gl}}{(H_d - H_\ell) - (H_g - H_{g\ell})}$$

Vapor compression with free expansion

This refrigeration cycle is shown in Figure 3.4.

Because of highly irreversible expansion in the valve, this cycle is less efficient than the turbine expansion but the turbine operating on a two-phase vapor-liquid mixture is expensive. Therefore in small plants, such as household refrigerators, the throttle valve is used because of its lower cost and simplicity. The throttle

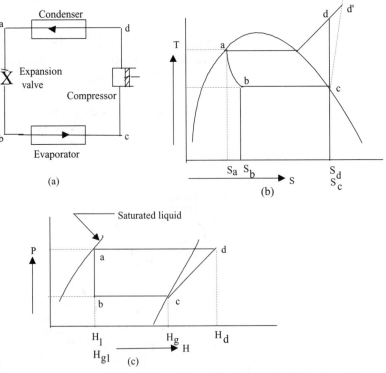

Figure 3.4 Vapor compression refrigeration cycle with free expansion. (a) System sketch (b) T-S diagram (c) P-H diagram.

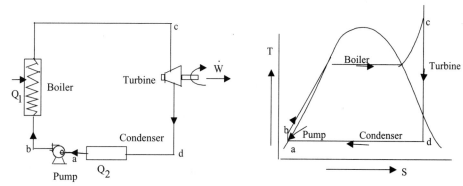

Figure 3.5 Rankine cycle without superheating (a) Rankine cycle (b) T-S diagram.

valve free expansion is isenthalpic. The coefficient of performance, β, of the vapor compression cycle with free expansion is given by

$$\text{Coefficient of performance, } \beta = \frac{H_g - H_\ell}{H_d - H_g}$$

Other calculations

Besides the coefficient of performance, it is required to know the rate of circulation of the refrigerant in order to design and size the compressor, condenser, refrigerator coils, and other auxiliary equipment. For this, the relations summarized below are useful.

$$\text{Net refrigeration effect} = \hat{H}_g - \hat{H}_{gl} = \hat{H}_g - \hat{H}_l$$

$$\text{Mass flow rate of refrigerant} = M = \frac{12{,}000 \; Btu/h \cdot ton}{Q_1 \; Btu/lb}$$

$$\text{Heat of compression} = \hat{H}_d - \hat{H}_g$$

$$\text{Work of compression} = -W_c = (\hat{H}_d - \hat{H}_g)\,M$$

$$\text{Condenser heat load} = Q_2 = \hat{H}_d - \hat{H}_\ell$$

$$\text{hp} = 42.4 \; Btu/min$$

$$\text{hp/ton} = \frac{\text{work of compression (Btu/min}\cdot\text{ton)}}{42.4 \; Btu/min} = \frac{4.713}{\beta}$$

Rankine Cycle

The Rankine cycle is shown in Figure 3.5. It is a power generation cycle and operates with steam as working fluid.

The efficiency of Rankine cycle is given by

$$\eta = \frac{(H_c - H_d) - (H_b - H_a)}{H_c - H_b}$$

HEATS OF MIXING AND ENTHALPY OF A SOLUTION

Heat of mixing is the enthalpy change on dissolving a solute in a solvent. Standard integral heat of solution is the change in enthalpy of the system when 1 mole of a solute is mixed with n_1 moles of a solvent at constant temperature of 25 °C and

1 atmosphere. The enthalpy is given by

$$H_s = n_1 \hat{H}_1 + n_2 \hat{H}_2 + n_2 \Delta \hat{H}_{s2}$$

where
H_s = enthalpy of $n_1 + n_2$ moles of solution of components 1 and 2 at temperature T relative to temperature T_{ref} (reference temperature)
$\hat{H}_1, \hat{H}_2$ = molar enthalpies of components 1 and 2 at temperature T relative to the reference temperature, T_{ref}
$\hat{H}_{s2}$ = integral heat of solution of component 2 at temperature T

GIBBS' PHASE RULE

For equilibrium states of multi-component multiphase systems, the variables are not all independent. In consequence, if a limited number of them are fixed, the remaining variables are automatically determined. Gibbs' phase rule allows us to get this number of independent variables. This number of independent variables is called the *degrees of freedom* of the system. The degrees of freedom are given by the phase rule as:

$$F = N - r + 2 - P$$

where
F = degrees of freedom
N = number of components
r = number of reactions
P = number of phases present

If the system does not involve any chemical reaction, the degrees of freedom are given by

$$F = N + 2 - P$$

REFERENCES

1. Das, D. K., and R. K. Prabhudesai, *Chemical Engineering License Review*, 2nd ed., Kaplan AEC Education, 2004.
2. Smith, J. M., and H. C. van Ness, *Introduction to Chemical Engineering Thermodynamics*, 6th ed., McGraw-Hill, 2000.
3. *AIChE Modular Instruction Series*, Thermodynamics, *vol 1–4*.

PROBLEMS

3.1 A piston cylinder initially contains 100 gmol of an ideal gas at a pressure of 516.3 kN/m^2 A and 10°C. External pressure is 101.3 kN/m^2. The piston is weighted with 100 kg of weight and held in place with latches. The temperature is kept constant. When latches are released, the piston moves up and comes to rest when the forces are balanced. Calculate the work done during this expansion. Next the mass is removed and the cylinder is allowed to come to equilibrium against the external pressure. What is the work done in this step? What would be the work done if the gas were expanded reversibly against the external pressure. Area of cross section of piston is 0.0029 m^2. Assume temperature is constant during both expansions. Report calculated work in units of kJ.

3.2 Two Carnot engines are operating in series. The first one absorbs heat at a temperature of 1111 K and rejects heat to the second engine at a temperature T. The second engine receives the heat at the intermediate temperature T and rejects it to a reservoir at 300 K. Calculate T if (a) the efficiencies of the two engines are equal and (b) the works done by the two engines are equal.

3.3 Express Van der Waal's equation of state as Virial equation.

3.4 Calculate the specific volume of CH_3Cl at 1379 kPa and 205 °C Using Van der Waal's equation. Van der Waal's constants for CH_3Cl are a = 757.4 kPa (m^3/kmol)^2 and b = 0.065 m^3/kg mol.

3.5 A Carnot engine operates in a closed system by absorbing heat at 927 °C and rejecting heat at 27 °C and produces 5021 kJ of net work. Determine the heat input and output of the engine.

3.6 The constant pressure specific heat of acetonitrile vapor at low pressures is given by the equation

$$C_p = 21.3 + 11.562 \times 10^{-2} T - 3.812 \times 10^{-5} T^2$$

where T is in K and C_P is in kJ/kg mol·K

Estimate specific heat ratio (C_P/C_V) for acetonitrile at 1000 K.

3.7 25 kg of carbon dioxide are to be heated from 300 to 700 K at constant volume. Calculate the number of kJ to be supplied. Specific heat of carbon dioxide is given by $C_P = 43.26 + 0.0115T$ where C_P is in kJ/kg mol · K and T is in degrees Kelvin.

3.8 Liquid allyl alcohol has a vapor pressure of 53.32 kPa at 80.2 °C and its normal boiling point is 96.6 °C. Calculate its heat of vaporization over the temperature range of 80.2 to 96.6 °C.

3.9 In a refrigeration cycle using HFC-134a, the refrigerant enters the expansion valve as saturated liquid at 10 barA and 40°C. The downstream pressure is 1.38 barA. What is the entropy increase (kJ/kg · K) in the expansion?

3.10 Using the following data, calculate the fugacity coefficient for ethane at 522 K and 69 bar pressure.

Temperature K	Pressure bar	Volume m³/kg	Enthalpy of vapor kJ/kg	Entropy of vapor kJ/kg·K
522	0.7	2.1	1562.1	8.971
522	69.0	0.0197	1520.0	7.633

3.11 The stored energy of a closed system increases by 110 kJ when work is done on the system in the amount of 200 kJ. Is the heat transferred to or from the system and how much?

SOLUTIONS

3.1 (a) Effective external pressure after latches are released and equilibrium is reached

$$= 101.3 + \frac{100 \times 9.81}{1000 \times 0.002} = 101.3 + 338.3 = 439.6 \text{ kN/m}^2$$

Assume ideal gas behavior

$$\text{Closed system } W = \int_{V_1}^{V_2} P_{ext} dV = P_{ext}(V_2 - V_1) = P_m(V_m - V_1)$$

Where m is used to indicate intermediate pressure and volume.

$$P_m = 439.6 \text{ kN/m}^2$$

Thus $W = P_m(V_m - V_1) = P_m V_m (1 - P_m/P_1)$ since $V_1/V_m = P_m/P_1$
$= nRT(1 - P_m/P_1)$ since $P_m V_m = nRT$
$= 100(8.314)(283)(1 - 439.6/516.3)$
$= 34955 \text{ J} = 34.955 \text{ kJ}.$

(b) In the second expansion, the gas expands against constant external pressure of 101.3 kN/m². Initial pressure is now 439.6 kN/m². Then

$$W = nRT(1 - P_1/P_m) - 100(8.314)(283)(1 - 101.3/439.6)$$
$$= 181.076 \text{ kJ}$$

Total work done = 34.955 + 181.076 = 216.03 kJ

(c) Reversible work is given by $W = \int_{V_1}^{V_2} P dV$ (reversible)

For ideal gas, at constant temperature $PdV = -VdP = RT\dfrac{dP}{P}$

Therefore $W = -RT \ln \dfrac{P_2}{P_1} = -8.314(283) \ln \dfrac{101.3}{516.3} = 3.832 \text{ kJ/g mol}$

Then work per 100 gmol = 100(3.832) = 383.2 kJ

3.2 (a) Carnot efficiency $= \dfrac{T_H - T_C}{T_H}$

Efficiencies of the two engines are equal. Hence $\dfrac{T_H - T}{T_H} = \dfrac{T - T_C}{T}$

Therefore, we can get from the equality $T/T_H = T_C/T$ or $T = \sqrt{T_H T_C}$, $T_H = 1111$ K and $T_C = 300$ K

Then $T = \sqrt{1111 \times 300} = 577.3$ K

(b) In this case works are equal. Therefore, $W_1 = W_H - Q_T = W_2 = Q_T - Q_C$

Dividing by Q_T, one obtains $\dfrac{Q_H}{Q_T} - 1 = 1 - \dfrac{Q_C}{Q_T}$ (equation 1)

Also $\dfrac{Q_H}{Q_T} = \dfrac{T_H}{T}$ and $\dfrac{Q_C}{Q_T} = \dfrac{T_C}{T}$

By substitution of these relations in equation 1, we get $\dfrac{T_H}{T} + \dfrac{T_C}{T} = 2$

which yields $T = \dfrac{T_H + T_C}{2} = \dfrac{1111 + 300}{2} = 705.\,K$

3.3 Van der Waal's equation is given by $\left(P + \dfrac{a}{\hat{V}^2}\right)(\hat{V} - b) = RT$ for 1 gmol

This can be expressed as $P = \dfrac{RT}{\hat{V} - b} - \dfrac{a}{\hat{V}^2}$

Now multiply each side by $\hat{V}$, then $P\hat{V} = \dfrac{RT\hat{V}}{\hat{V} - b} - \dfrac{a\hat{V}}{\hat{V}^2}$

Which can be rearranged as $P\hat{V} = RT(1 - b/\hat{V})^{-1} - \dfrac{a}{\hat{V}}$

Now expand the term in braces using the binomial theorem as follows

$$\left(1 - \dfrac{b}{\hat{V}}\right)^{-1} = 1 + \dfrac{b}{\hat{V}} + \dfrac{b^2}{\hat{V}^2} + \dfrac{b^3}{\hat{V}^3} + \cdots$$

Substituting this expansion in the equation and collecting together the terms with the same power of $\bar{V}$, the virial form of Van der Waal's equation results as follows:

$$P\hat{V} = 1 + \dfrac{RT(b - a)}{\hat{V}} + \dfrac{RTb^2}{\hat{V}^2} + \dfrac{RTb^3}{\hat{V}^3} + \cdots$$

3.4 Van der Waal's equation $\left(P + \dfrac{a}{v^2}\right)(v - b) = RT$ wherein specific volume $= v = \bar{v}$

Simplifying the equation, $pv - Pb + \dfrac{a}{v} - \dfrac{ab}{v^2} = RT$

Multiply each term by v^2, transpose all terms to the left hand side, and divide by P to get the equation

$$v^3 - \left(\dfrac{Pb + RT}{P}\right)v^2 + \dfrac{a}{p}(v - b) = 0$$

$P = 1379$ kPa $R = 8.314 \dfrac{kPa \cdot m^3}{kg\,mol \cdot K}$

$a = 757.4$ kPa $b = 0.065$ m³/kg mol

Substitution in the equation results in the following relation:
$$v^3 - 2.95v^2 + 0.55\,(v - 0.065) = 0$$

Which is cubical in v and must be solved by trial or graphically.

The calculations are done as follows:
(Hint: Assume the first trial value obtained from the ideal gas law).

Assumed value of v	2.9	2.62	2.76
Calculated value of LHS	1.105	–0.87	0.004436 (close to 0)

Therefore $v = 2.76$ m³/kg mol

3.5 A Carnot engine is a reversible engine. Therefore, $\dfrac{Q_H}{T_H} = \dfrac{Q_C}{T_C}$

Also $W = Q_H - Q_C = 5021$ kJ

$T_H = 1200$ K $T_C = 300$ K

Then $Q_H = Q_C (T_H/T_C) = Q_C (1200/300) = 4Q_C$

Hence $Q_H - Q_C = 4Q_C - Q_C - 5021$ kJ

From which $Q_C = 5021/3 = 1673.7$ kJ

Therefore $Q_H = 4 \times 1673.7 = 6695$ kJ

The engine absorbs 6695 kJ at 1200 K and rejects 1673.7 kJ at 300 K.

3.6 $T = 1000$ K

At this temperature, assumption of ideal gas behavior is appropriate, and therefore,

$$C_P - C_V = R$$
$$C_p = 21.3 + 11.562 \times 10^{-2} T - 3.812 \times 10^{-5} T^2$$
$$= 21.3 + 11.562 \times 10^{-2} (1000) - 3.812 \times 10^{-5} (1000)^2$$
$$= 98.8 \text{ kJ/kg mol}$$
$$C_V = 98.8 - 8.314 = 90.94 \text{ kJ/kg mol} \cdot \text{K}$$
$$C_P/C_V = 98.8/90.49 = 1.092$$

3.7 Assume CO_2 to behave ideally over the temperature range of the problem
Therefore

$$C_V = C_P - R = 43.26 + 0.0115T - 8.314$$
$$= 34.946 + 0.0115\, T \quad T \text{ in Kelvin.}$$

From first law, $U = Q = C_V dT$

$$T_2 = 700 \text{ K}$$
$$T_1 = 300 \text{ K}$$

Then $Q = \int_{T_1}^{T_2} C_V dT = \int_{300}^{700} (34.946 + 0.0115T) dT$

$$= 34.946(T_2 - T_1) + \dfrac{0.0115}{2}\left(T_2^2 - T_1^2\right)$$
$$= 34.946 (700 - 300) + 0.00575 (700^2 - 300^2)$$
$$= 16278.4 \text{ kJ/kg mol}$$

kmol of CO_2 to be heated $= 25/44 = 0.5682$ kg mol

Heat to be supplied $= 0.5682 (16278.4) = 9249.4$ kJ

3.8 Assume H = constant and use the equation

$$\ln \frac{P_2}{P_1} = \frac{\Delta H_V}{R}\left(\frac{1}{T_1} - \frac{1}{T_2}\right)$$

$$T_1 = 80.2 + 273 = 353.2 \text{ K} \quad T_2 = 96.6 + 273 = 369.6 \text{ K}$$

Then $\ln \dfrac{101.3}{53.32} = \dfrac{\Delta H_V}{8.314}\left(\dfrac{1}{353.2} - \dfrac{1}{396.6}\right)$

from which $\Delta H_V = 42472$ kJ/kg mol

3.9 The expansion through an expansion valve is isenthalpic. Therefore $h_i = h_o$

From P-H diagram of HFC–134a (SI system)

For liquid, $h_i = 257$ kJ/kg $S_i = 1.19$ kJ/kg·K

At 1.38 barA, $h_o = 257$ kJ/kg of liquid-vapor mixture

Following constant enthalpy line at 1.38 barA, $S_o = 1.225$ kJ/kg · K

The entropy increase during expansion = $S_o - S_i = 1.225 - 1.19 = 0.035$ kJ/kg · K

3.10 $$dG_T = RTd\ln f \quad \text{or} \quad d\ln f = \frac{dG_T}{RT}$$

Integration between high pressure P and low pressure at constant temperature gives

$$RT \ln \frac{f}{f^*} = \hat{G} - \hat{G}^* \quad \text{and since } G = H - T$$

$$\ln \frac{f}{f^*} = \frac{1}{R}\left[\frac{\hat{H} - \hat{H}^*}{T} - (\hat{S} - \hat{S}^*)\right]$$

If the reference state is a low pressure, then $f^* = P^*$ and then

$$\ln \frac{f}{P^*} = \frac{1}{R}\left[\frac{\hat{H} - \hat{H}^*}{T} - (\hat{S} - \hat{S}^*)\right]$$

Assume ethane behaves as an ideal gas at 0.7 barA and 522 K. Then

$$P^* = 0.7 \text{ bar A}, \quad \hat{H}^* = 1562.1 \text{ kJ/kg}, \quad \hat{S}^* = 8.971 \text{ kJ/kg·K}$$

At 69 barA and 522 K, $\hat{H}$ = 1520 kJ/kg and $\hat{S}$ = 7.633 kJ/kg·K

Therefore, $\ln \dfrac{f}{P^*} = \dfrac{30}{8.314}\left[\dfrac{1520-1562.1}{522} - (7.633-8.971)\right] = 4.533$

Then $\dfrac{f}{P^*} = e^{4.533}$ 93.04 barA and therefore $f = 93.04(0.7) = 65.1$ barA

Fugacity coefficient = $\phi = \dfrac{f}{P} = 65.1/69 = 0.9435$

3.11 Energy balance on the closed system gives

$$(U+PE+KE)_E - (U+PE+KE)_B = (H+PE+KE)_I - (H+PE+KE)_O + Q + W$$
$$111122$$

1   Negligible potential and kinetic energy effects

2   Closed system, therefore no mass transfer across boundaries of the system

Thus energy balance reduces to $U_E - U_B = Q + W$

Or in difference form $\Delta U = Q + W$

ΔU = +110 kJ   W = + 200 kJ since work is done on the system.

Inserting in the final energy balance

$$110 = Q - (+200)$$

and Q = –90 kJ

Since Q is negative, 90 kJ must be removed from the system.

(Note: The above solution is given at length for the sake of clarity. One should be able to write the final energy balance by working out the reasoning mentally to save considerable time. One should write the energy balance $\Delta U = Q + W$ at once and proceed further with the solution in this example.)

CHAPTER 4

Mass Transfer

OUTLINE

FICK'S LAW 55
Estimation of Binary Diffusivities ■ Diffusivity
in a Multi-Component Mixture

MASS TRANSFER UNDER NON-FLOW CONDITIONS 57

MASS TRANSFER COEFFICIENTS 58

MASS TRANSFER FROM A GAS INTO A
FALLING LIQUID FILM 59

MASS TRANSFER FROM SPHERES 59

TURBULENT MASS TRANSFER 60
Film Model ■ Higbie Penetration Model ■ Surface Renewal
Theory of Danckwerts ■ Turbulent Boundary Layer Theory

INTERPHASE MASS TRANSFER 61

OVERALL MASS TRANSFER COEFFICIENTS 62

MASS TRANSFER IN PACKED BEDS 63

DIFFUSION IN SOLIDS 65

REFERENCES 66

Mass transfer involves movement of molecules under some driving force such as concentration or temperature difference. However, mass transfer discussed here will primarily deal with concentration difference. Mass transfer takes place by two mechanisms: (1) molecular diffusion and (2) convective diffusion.

FICK'S LAW

This basic relation for mass transfer by molecular diffusion in a binary mixture of A and B is given by

$$J_A = -D_{AB} \frac{dC_A}{dz}$$

where
- J_A = molar flux of component A along axis z and moving at molar average velocity
- D_{AB} = diffusivity of A in the solution containing A and B
- C_A = concentration of i
- Z = distance coordinate in the direction of molecular diffusion

Another expression for the same case is

$$J_A = -c_m D_{AB} \frac{dy_A}{dz}$$

where
- y_A = mol fraction of A
- c_m = average molar density of the mixture

Estimation of Binary Diffusivities

For gases, an empirical equation to calculate diffusivity of a component in a binary mixture is

$$D_{AB} = \frac{0.001 T^{1.75}(1/M_A + 1/M_B)^{1/2}}{P\left[(\Sigma v)_A^{1/3} + (\Sigma)_B^{1/3}\right]^2} \text{ cm}^2/\text{s}$$

where
  v is the atomic diffusion volume of a simple molecule
  Other symbols have usual meaning, e.g., p = pressure
  Σv is obtained by summation of the atomic diffusion volumes

The theoretical equation for calculation of diffusivity in binary gas mixtures at low pressures is

$$D_{AB} \frac{0.0018587 T^{3/2}(1/MA + 1/M_B)^{1/2}}{P\sigma_{AB}^2 \Omega_D} \text{ cm}^2/\text{s}$$

where
  T = temperature, K
  M_A and M_B = molecular weights
  P = Pressure, atm
  $\Omega_D = f(kT/\epsilon_{AB})$, collision integral
  k = Boltzmann's constant
  $\epsilon_{AB}, \sigma_{AB}$ Lennard-Jones force constants for the binary

The values of ϵ_{AB} and σ_{AB} are obtained from values for the pure components by relations

$$\frac{\epsilon_{AB}}{k} = \left(\frac{\epsilon_A}{k}\frac{\epsilon_B}{k}\right)^{1/2} \quad \text{and} \quad \sigma_{AB} = \frac{1}{2}(\sigma_A + \sigma_B)$$

Values of ϵ/k and σ are available in the literature. If not, they may be estimated by the following relations:

$$\frac{\epsilon}{k} = 0.75 T_c \quad \text{and} \quad \sigma = \frac{5}{6} V_c^{1/3}$$

where

T_c = critical temperature, K
V_c = critical volume, cm^3/gmol
σ = Lennard-Jones potential parameter, A°

For liquids, the Wilke and Chang relation is given by

$$D_{AB}^o = 7.4 \times 10^{-8} \left[(\phi M_B)^{1/2} \frac{T}{\mu_B V_A^{0.6}} \right]$$

where

ϕ = association parameter of solvent B
ϕ = 2.26 for water as solvent, 1.9 for methanol, 1.5 for ethanol, and 1 for unassociated solvents such as benzene and ethyl ether
μ_B = viscosity of B, cP (centipoises)
V_B = molar volume of solute A at its normal boiling point, cm^3/gmol

Diffusivity in a Multi-Component Mixture

The diffusivity of a component in a mixture is expressed by the Stefan-Maxwell equation. A simplified version of the equation is

$$D_{1-mix} = \frac{1}{\sum_{i=2}^{n} y_i'/D_{1-i}} \quad \text{where} \quad y_i' = y_i \Big/ \sum_{j=2}^{n} y_j$$

y_i' = mol fraction of component i, calculated by excluding component 1

MASS TRANSFER UNDER NON-FLOW CONDITIONS

Some important relations for mass transfer under non-flow conditions are as follows:

Steady state diffusions of A into stagnant gas-film of B (gas):

$$N_{Az} \frac{D_{AB} P}{RT(z_2 - z_1)} \frac{1}{(p_B)_{lm}} (p_{A1} - p_{A2}) = \frac{c D_{AB}}{(z_2 - z_1)} \ln \frac{x_{B2}}{x_{B1}}$$

where

$$p_{lm} = \frac{p_{B2} - p_{B1}}{\ln(p_{B2}/p_{B1})}$$

Steady state equimolar counter diffusion of A and B (gas):

$$N_{Az} = \frac{D_{AB}}{z_2 - z_1} (C_{A1} - C_{A2}) = \frac{D_{AB}}{RT(z_2 - z_1)} (p_{A1} - p_{A2})$$

Steady state diffusion of A into stagnant liquid film of B:

$$N_{Az} = \frac{D_{AB} \rho}{M_{av}(z_2 - z_1)} \frac{1}{(x_B)_{lm}} (x_{A1} - x_{A2})$$

Equimolar counter diffusion (liquids):

$$N_{Az} = \frac{D_{AB}}{(z_2 - z_1)}\left(\frac{\rho}{M}\right)_{av} (x_{A1} - x_{A2})$$

Vaporization of a spherical drop of a liquid:

$$N_{Az} = \frac{\rho D_{AB}}{r_0} \frac{1}{(W_B)_{lm}} (W_{B\delta} - W_{BS})$$

MASS TRANSFER COEFFICIENTS

Mass transfer coefficient is defined by the relation

Mass flux = (mass transfer coefficient)(driving force)

For example, for the diffusion of A through non-diffusing B

$$N_A = k_G(p_{A1} - p_{A2}) = k_y(y_{A1} - y_{A2}) = k_c(C_{A1} - C_{A2}) \quad \text{for gases}$$
$$N_A = k_x(x_{A1} - x_{A2}) = k_L(C_{A1} - C_{A2}) \quad \text{for liquids}$$

Units of mass transfer coefficients will depend upon the units chosen for the driving force viz. the concentration differences, partial pressures, or mol fractions.

For gases, $k_g = (k_y/P_t) = (k_c/RT)$ since $p_A = y_A P_t$ and $C_A = (p_A/RT)$

In like manner, for liquids, the relationship is $k_L C = k_x$ since $C_A = Cx_A$, where k is a mass transfer coefficient in appropriate units. For example, $k_y = $ mols/(cm$^2 \cdot$ s $\cdot \Delta y$). Mass transfer coefficients for various situations are correlated in terms of dimensionless numbers as in heat transfer. These are

$$\text{Sherwood number} = \left(\frac{kD}{D_{AB}}\right) = Sh$$

$$\text{Reynold's number} = \left(\frac{Du\rho}{\mu}\right) = Re$$

$$\text{Schmidt number} = \left(\frac{\mu}{\rho D_{AB}}\right) = Sc$$

$$\text{Grashof number} = \left(\frac{gD^3 \rho \Delta \rho}{\mu^2}\right) = Gr$$

The product of Re and Sc is called the Peclet number, Pe, and St, the Stanton number, is the ratio Sh/Pe (Sherwood number/Peclet number).

Depending upon the nature of the mass transfer coefficient used, expressions for the Sherwood number will change. Thus the Sherwood number is

$$Sh = \frac{Fl}{cD_{AB}} = \frac{k_G p_B, \overline{M}^{RTl}}{D_{AB} P_t} = \frac{k_c p_B, \overline{M}^l}{P_t D_{AB}}$$

MASS TRANSFER FROM A GAS INTO A FALLING LIQUID FILM

For $Re < 100$, $k_{L,av} = 3.41 \dfrac{D_{AB}}{\delta}$ or $Sh_{av} = \dfrac{k_{l,av}\delta}{D_{AB}} = 3.41$

where δ = film thickness

For $Re > 100$ $\quad k_{L,av} = \left(\dfrac{6 D_{AB} \Gamma}{\pi \rho \delta L}\right)^{1/2}$ and $\quad Sh_{av} = \left(\dfrac{3}{2\pi}\dfrac{\delta}{L} Re\, Sc\right)^{1/2}$

where
 δ = film thickness
 Γ = mass rate of liquid flow per unit width of film in the x direction
 L = length in the z direction

For a falling film, the film thickness is given by

$$\delta = \left(\dfrac{3\mu\Gamma}{\rho^2 g}\right)^{1/3} = \left(\dfrac{3 u \bar{y} \mu}{\rho g}\right)^{1/2}$$

$k_{L,av}$ is given by: $\quad k_{L,av} = \dfrac{u_{\bar{y}}\delta}{L} \ln \dfrac{C_{Ai} - \overline{C}_A}{C_{A,i} - \overline{C}_{A,L}}$

Then flux, $N_{A,av} = k_{L,av}(C_{A,i} - \overline{C}_A)_M$

where $(C_{A,i} - \overline{C}_{A,L})$ is the logarithmic average concentration difference over length L

MASS TRANSFER FROM SPHERES

For stationary fluid

$$Sh_0 = \dfrac{k_c D}{D_{AB}} = 2$$

where
 N_{sh} = Sherwood number for stationary fluid

For convection

$$Sh = Sh_0 + 0.6\, Re_{0.5}\, Sc_{1/3} \qquad \text{[Frossling Equation]}$$

Another equation for convective mass transfer is

$$\dfrac{k_c D}{D_{AB}} = 2 + 1.0\, Re^{1/3} Sc^{1/3} = 2 + 1.0\, Pe^{1/3} \quad \text{for } Pe > 1.0$$

Mass transfer coefficients for other geometries, such as from flat plates, in laminar flow through tubes, or rotating discs, are available in the literature. Table 4.1 summarizes some of these.

Chapter 4 Mass Transfer

Table 4.1 Mass transfer coefficients correlations for turbulent flow

Motion of Fluid	Range of Condition	Equation
Inside tube (circular)	$Re = 4000–60000$	$j_D = 0.023\, Re^{-0.17}$
	$Sc = 0.6–3000$	$Sh = 0.023\, Re^{0.83} Sc^{1/3}$
Flow parallel to flat plate (unconfined)	$Re = 10000–400000$	$j_D = 0.0149\, Re^{-0.12}$
	$Sc > 100$	$Sh = 0.0149\, Re^{0.88} Sc^{1/}$
Liquid film in wetted-wall tower	$4\Gamma/\mu = 0–1200$	See equations for laminar flow
Transfer between liquid and gas	$4\Gamma/\mu = 1300–8300$	$Sh = 1.76 \times 10^{-5}\left(\dfrac{4\Gamma}{\mu}\right)^{1.506} Sc^{0.5}$
	$Re = 4000–30000$	$\dfrac{k_c d}{D_{AB}}\dfrac{P_{BM}}{p} = 0.023\, Re^{0.83} Sc^{0.44}$
		or $\dfrac{k_c d}{D_{AB}}\dfrac{P_{BM}}{p} = 0.0328\, Re^{1.1} Sc^{1/3}$
Perpendicular to single cylinders	$Re = 400–25000$	Gases: $Sh = 0.28 Re^{0.5} Sc^{0.44}$
		Liquids: $Sh = 0.281 Re^{0.6} Sc^{1/3}$
Flow past single spheres	$Re = 2000–17000$	$Sh = 0.347\, Re^{0.62} Sc^{1/3}$

TURBULENT MASS TRANSFER

Most practical situations of mass transfer involve turbulent flow. In these situations, it is not possible to describe the flow conditions adequately in mathematical terms. Therefore, one relies mainly on experimental data. But these are limited in scope with respect to fluid conditions and the range of its properties. In order to extend applicability of experimental data, many theories or models have been developed. Other methods involve use of analogy between mass, heat, and momentum transfer to develop correlations for mass transfer in turbulent flow. Some of these models will be very briefly reviewed next.

Film Model

The film model assumes the resistance to mass transfer resides entirely in a stagnant thin film and the concentration difference lies entirely in this film. It also assumes that the concentration difference is only due to molecular diffusion. This model is based on the film model used for convective heat transfer. The mass flux through this film is given by

$$N = -D\left(\frac{dc}{dx}\right)_{x=0} = \frac{D}{\delta}(C_o - C_L) = \frac{D}{\delta}\Delta C = k_L(\Delta C)$$

Higbie Penetration Model

The film model assumes no accumulation of diffusing species in the film and a steady state is assumed for transfer across the film. Higbie pointed out that in industrial gas-liquid contacting equipment, contact times are very small, which prevents attainment of steady state. He developed his penetration model to take into account the transient nature of solute diffusion. Based on his derivation, the average flux and time average mass-transfer coefficient are given by

$$N_{A,av} = 2(C_{Ai} - C_o)\left(\frac{D_{AB}}{\pi t}\right)^{1/2}$$

and $\quad k_L = 2(D_{AB'}/\pi t)^{1/2}, \quad$ also $Sh = 1.13 \, Re^{0.5} Sc^{0.5}$

Surface Renewal Theory of Danckwerts

Important improvement in the penetration model was proposed by Danckwerts. Higbie assumed equal exposure time for repeated contacts of the fluid with the interface. Danckwerts used a wide range of contact times and averaged the various degrees of penetration. He derived the following expressions for mass flux and the mass transfer coefficient.

$N_A = (C_{Ai} - C_o)\sqrt{Ds}$ where s is fractional surface renewal rate. $\quad k_c = \sqrt{Ds}$

Toor-Marchellow/Dobbins models are also available.

Turbulent Boundary Layer Theory

This theory explains the mass transfer between a fixed surface and a turbulent stream of fluid. There is no slip at the wall and the turbulence is suppressed in the fluid in contact with the wall. At the wall the transfer is by molecular diffusion, whereas eddy diffusion and not molecular diffusion prevails in the turbulent stream far from the wall. In between these two positions both molecular and eddy diffusion contribute to mass transfer.

As in heat and momentum transfer, turbulent mass transfer flux can be written as the sum of two fluxes, one due to turbulence and the other due to laminar boundary as follows:

$$(N_A)_{Total} = -(D_{AB} + \varepsilon_D)\frac{dC_A}{dy}$$

where ε_D = eddy mass diffusivity

With the use of boundary layer theory, an expression for the mass transfer coefficient has been developed for turbulent mass transfer. The expressions derived are

$$k = 0.664 \, Re^{0.5} Sc^{0.5} \frac{D_{AB}}{L} \quad \text{and} \quad k_L = \frac{D_{AB}}{L}(0.0365)(0.302) Re^{1.08} Sc^{7/15}$$

Analogy correlations (heat transfer and momentum transfer) and *J* factor methods are also applied to determine mass transfer coefficients.

INTERPHASE MASS TRANSFER

Lewis and Whitman assumed that in mass transfer between two fluid phases, resistances to mass transfer occur only in the fluids and there is no resistance to

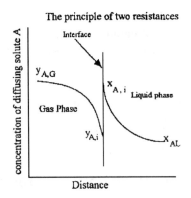
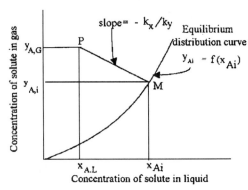

Figure 4.1 Mass transfer between two phases

solute transfer across the interface between the two phases. This is called the principle of two resistances (Figure 4.1).

At the interface, $y_{A,il}$, and x_{Ail} are in equilibrium. If k_x and k_y are local mass transfer coefficients, $N_A = k_y(y_{AG} - y_{Ai}) = k_x(x_{Ai} - x_{AL})$, where $(y_{AG} - y_{Al})$ and $(x_{Ai} - x_{Al})$ are the driving forces for mass transfer in the gas and the liquid phases. A rearrangement gives the relation

$$\left(\frac{y_{Ag} - y_{Ai}}{x_{AL} - x_{Al}}\right) = -\frac{k_x}{k_y}$$

Thus if the mass transfer coefficients are known, the interfacial concentrations and the flux N_A can be determined graphically or analytically by solving the above equation in conjunction with the equilibrium curve.

OVERALL MASS TRANSFER COEFFICIENTS

As in heat transfer, in mass transfer also, one can write overall mass transfer coefficients in terms of the individual mass transfer coefficients in terms of the bulk concentrations. With liquid phase as basis, it is possible to derive the following relation:

$$\frac{1}{K_x} = \frac{1}{mk_y} + \frac{1}{k_x}$$

If the gas phase is chosen as the basis to derive the overall mass transfer coefficients, the relation is

$$\frac{1}{K_y} = \frac{1}{k_y} + \frac{m}{k_x}$$

where m is the slope of the equilibrium curve assumed to be a straight line
It should be noted that if the gas phase resistance controls

$$\frac{1}{K_y} \approx \frac{1}{k_y}$$

and if the liquid phase resistance controls

$$\frac{1}{K_x} \approx \frac{1}{k_x}$$

These relations imply that

$$y_{Ag} - y^*_{Ai} \not\subset x_{Ai} - y_{Ai} \text{ for gas phase controlling}$$
$$x^*_A - x_{AL} \not\subset x_{Ai} - x_A \text{ for liquid phase controlling}$$

In practice, average mass transfer coefficients are used, and these are correlated by dimensionless numbers.

For general cases where (1) mass transfer rates are high, (2) diffusion of more than one substance is involved, or (3) equimolar counter-diffusion is not involved F_G and F_L should be used. Thus mass-transfer flux is given by

$$N_A = \frac{N_A}{\Sigma N} F_G \ln \frac{N_{A'}\Sigma N - y_{Ai}}{N_{A'}\Sigma N - y_{Ag}} = \frac{N_A}{\Sigma N} F_L \ln \frac{N_{A'}\Sigma N - x_{AL}}{N_{A'}\Sigma N - x_{Ai}}$$

where F_G and F_L are generalized gas and liquid phase mass transfer coefficients for component A

One can also define generalized overall mass transfer coefficients F_{OG} and F_{OL}.

The following relations can be derived for two simple cases.

1. Diffusion of one component:

$$eN_{A'}F_{OG} = eN_{A'}F_G = m'\frac{1-x_{A,L}}{1-y_{A,G}}\left(1 - e^{-N_{A'}F_L}\right)$$

and $\quad e^{-N_{A'}F_{OL}} = \frac{1}{m''}\frac{1-y_{AG}}{1-x_{Al}}\left(1 - e^{N_{A'}F_G}\right) + e^{-N_{A'}F}$

2. Equimolar counter-diffusion:

$$\frac{1}{F_{OG}} = \frac{1}{F_G} + \frac{m'}{F_L} \quad \text{and} \quad \frac{1}{F_{OL}} = \frac{1}{m''F_G} + \frac{1}{F_G}$$

where m and m are slopes of chords of the equilibrium curve and are dimensionless

MASS TRANSFER IN PACKED BEDS

Mass transfer in packed beds takes place between two flowing phases. The usual approach is to determine the product of the mass transfer coefficient and the interfacial area per unit volume of packed bed. If a is the interfacial area per unit volume of packed bed, $a = A_i/V_t$, where A = total interfacial area in a packed tower of volume V_t. Then molar flow rate W_A of the diffusing species A is given by

$$W_A = k_z a V_t (Z_{AS} - Z_A) \quad \text{where } k_z \text{ can be either} \quad k_x a \quad \text{or} \quad k_y a.$$

However, mass transfer coefficients in packed towers are correlated in terms of a height of a transfer unit, H. The height of the transfer unit is defined as follows:

$$H_G = \frac{G}{k_y a}$$

where
  G = moles of gas/(unit time. unit cross section of empty tower)

Similarly,

$$H_L = \frac{L}{x_y a}$$

where
  L = moles of liquid/(unit time·unit cross section of empty tower).

H can also be defined to include the overall mass transfer coefficient as follows:

$$H_{OG} = \frac{G}{K_y a} \quad \text{(Overall height of transfer unit based on gas phase)}$$

$$H_{OL} = \frac{L}{K_x a} \quad \text{(Overall height of transfer unit based on liquid phase)}$$

Analytical expressions for H_{OG} and H_{OL}:

$$H_{OG} = \frac{G}{K_y a(1-y)_{lm}} = \frac{G}{K_{Ga} P_t (1-y)_{lm}} \quad \text{where} \quad (1-y)_{lm} = \frac{(1-y^*)-(1-y)}{\ln \frac{(1-y^*)}{(1-y)}}$$

$$H_{OL} = \frac{L}{K_x a(1-x)_{lm}} = \frac{L}{K_L aC(1-x)_{lm}} = \frac{LM_L}{K_L a\rho_L (1-x)_{lm}}$$

$$\text{where} \quad (1-x)_{lm} = \frac{(1-x)-(1-x^*)}{\ln \frac{1-x}{1-x^*}}$$

M_L = Molecular weight of liquid solvent

The overall mass transfer coefficients $K_G a$ and $K_y a$ can be calculated from the individual gas and liquid phase mass transfer coefficients and equilibrium relationship. Table 4.2 will be useful to identify correct units to be used in a particular case.

Liquid phase mass transfer coefficients may also be expressed in different units as follows:

(a) $\dfrac{1}{K_L a} = \dfrac{1}{k_L a} + \dfrac{m'_c}{k_G a}$ (b) $\dfrac{1}{K_L a} = \dfrac{1}{k_L a} + \dfrac{1}{m_c k_G a}$ (c) $\dfrac{1}{K_L a} = \dfrac{1}{k_L a} + \dfrac{\rho_m}{m k_G a p_t}$

(d) $\dfrac{1}{K_L a} = \dfrac{1}{k_l a} + \dfrac{\rho m}{m_x k_G a}$ (e) $K_L a = K_x a/C$ (f) $k_L a = k_x a/C$

By analogy with H_{OL} and H_{OG}, the gas film and liquid film transfer units can also be defined as follows:

(a) $H_{tG} = \dfrac{G}{k_y a(1-y)_{lm}} = \dfrac{G}{k_y a}$ (b) $H_{tG} = \dfrac{G}{k_G a P_t(1-y)_{lm}} \approx \dfrac{G}{k_G a P_t}$

(c) $H_{tL} = \dfrac{L}{k_x a(1-x)_{lm}} \approx \dfrac{L}{k_x a}$ (d) $H_{tL} = \dfrac{L}{k_L aC(1-x)_{lm}} \approx \dfrac{L}{k_L aC}$

Table 4.2 Units of mass transfer coefficients

(a) $\dfrac{1}{K_G a} = \dfrac{1}{k_G a} + \dfrac{m_c}{k_L a}$ — Here m_c is given by the equilibrium relationship $p^* = m_c C$ where p^* is in atm, and i in lbmol/ft^3

(b) $\dfrac{1}{K_G a} = \dfrac{1}{k_G a} + \dfrac{1}{m'_c k_L a}$ — In this equation, m_C is given by $p^* = C/m'_C$

(c) $\dfrac{1}{K_G a} = \dfrac{1}{k_G a} + \dfrac{m p_t}{k_L a \rho m}$ — where m is given by $y^* = mx$ and $\rho_m = \dfrac{\text{density of solution, lb/ft}^3}{\text{molecular weight of solution}}$

(d) $\dfrac{1}{K_G a} = \dfrac{1}{k_G a} + \dfrac{mx}{k_L a \rho m}$ — Where m_x is given by $\rho_M p^* = m_x x$

(e) $\dfrac{1}{K_G a} = \dfrac{1}{k_G a} + \dfrac{m}{k_x a}$ — Where m is given by $y^* = mx$

(f) $K_G a = K_y a / pt$

(g) $k_G a = k_y a / pt$

DIFFUSION IN SOLIDS

Fick's law can be applied for diffusion in solids under the following conditions: (1) concentration gradient is independent of time, (2) diffusivity is constant and independent of concentration, and (3) there is no bulk flow. Then, the flux is given by $N_A = -D_A dC/dZ$ where D_A is the diffusivity of A through the solid. For constant D_A, the following relations are obtained:

Diffusion through a flat slab: $N_A = \dfrac{D_A (C_{A1} - C_{A2})}{Z}$

Other solid shapes: $w = N_A S_{av} = D_A S_{av} (C_{A1} - C_{A2}) / z$

where S_{av} = average cross section for diffusion

Radial diffusion through a solid cylinder of inner and outer radii r_1 and r_2 and of length L

$$S_{av} = \dfrac{2\pi L (r_2 - r_1)}{\ln(r_2/r_1)} \quad \text{and} \quad Z = r_2 - r_1$$

Radial diffusion through a spherical shell: with inner and outer radii r_1 and r_2

$$S_{av} = 4\pi r_1 r_2 \quad \text{and} \quad z = r_2 - r_1$$

Diffusion in porous solids: Effective diffusivities are to be used in place of diffusivity in Fick's law equation.

Diffusion through polymers: The diffusional flux is expressed in terms of permeability P by the relation

$$V_A = \dfrac{D_A S_A (\bar{p}_{a_1} - \bar{p}_{a_2})}{Z}$$

where
- V_A = diffusional flux, cm³ gas (STP)/(cm²·s)
- D_A = diffusivity of A, cm²/s
- S_A = solubility coefficient or Henry's law constant, cm³ gas (STP)/(cm³ solid)·cmHg
- Z = thickness of polymeric membrane, cm
- $P = D_A S_A$ where P = permeability, cm³ gas (STP)/(cm²·s)(cmHg/cm)

REFERENCES

1. AIChE Modular Instruction Series, *Mass Transfer*, vol 1–4.
2. Das, D. K., and R. K. Prabhudesai, *Chemical Engineering License Review*, 2nd ed., Kaplan AEC, 2004.
3. Greencorn R. A., and D. P. Kessler, *Transfer Operations*, New York, McGraw-Hill, 1972.
4. Perry R. H., and D. W. Green, *Perry's Handbook for Chem. Engineers.*, Platinum ed., McGraw-Hill, 1999
5. Sherwood T. K., R. L. Pigford, and C. R. Wilke, *Mass Transfer*, McGraw-Hill, (1975).
6. Treybal R. E., *Mass Transfer Operations*, 3rd ed., McGraw-Hill, 1980.

PROBLEMS

4.1 Calculate the rate of diffusion of NaCl across a film of water (non-diffusing) solution 1.5 mm thick at 18°C when the concentrations on opposite sides of the film are 24 and 4 wt % NaCl, respectively. The diffusivity of NaCl is 1.3×10^{-5} cm^2/s at 18°C. Densities of 24 and 4 wt % NaCl at 18°C are 1181 and 1027 kg · m^3.

4.2 A rectangular tank 0.1 m^2 in cross section and 15 cm depth is filled with ethyl alcohol solution in water at 10°C to within 2 cm of the top. A fan is blowing air at 10°C and 1 bar over the tank. Alcohol concentration at the interface is 0.55 mole fraction. The diffusivity of ethyl alcohol is 0.039 m^2/h. How much alcohol will be lost in one day by diffusion? Assume that alcohol level is maintained constant, and local atm Pressure is 1 bar A.

4.3 Air is flowing in a tube whose inside surface is wetted with water. The temperature is 21°C. Calculate the flux of water evaporated from the pipe wall at a point where y_A, the bulk concentration (in mole fraction) of H$_2$O, is 0.0015. Additional data are as follows: ID of tube = 0.102 m, air velocity = 15.3 m/s, pressure = 1 barA., D_{AB} = 0.072 m^2/h, μ of air at 21°C = 0.018 mPa·s, density of air = 1.2 kg/m^3 · Sh = $0.023 Re^{0.83} Sc^{0.33}$, $N_A = K_G (P_{A1} - P_{A2})$, $K_G = K_C/(RT)$.

4.4 Hydrogen gas at 15.0 bar A and 30°C is transported through a steel pipe with ID and OD of 58.5 and 89 mm, respectively, to a reaction system. Henry's law constant for the solubility of hydrogen in steel is reported to be 1.67 bar·m^3/kg mol. The diffusion coefficient of hydrogen in steel is 0.3×10^{-12} m^2/s.

Calculate the mass flux of loss hydrogen by diffusion per hour per 100 meter length of pipe in units of kg/h.

SOLUTIONS

4.1 $Z = .0015$ m, $M_A = 58.5$, $M_B = 18.02$,
$D_{AB} = 1.3 \times 10^{-5}$ cm^2/s $= 1.3 \times 10^{-9}$ m^2/s
Density of 24% solution is 1081 kg/m^3. Therefore,

$$x_{A1} = \frac{0.24/58.5}{0.24/58.5 + 0.76/18.02} = \frac{0.0041}{0.054} = 0.08$$

$$x_{B1} = 1 - x_{A1} = 1.0 - 0.0886 = 0.9114$$

$$M = \frac{1}{0.0463} = 21.6 \text{ kg/kg mol} \qquad \frac{\rho}{M} = \frac{1081}{21.6} = 50.05 \text{ kg mol/m}^3$$

Similarly for 4% solution

$$x_{A2} = \frac{0.04/58.5}{0.04/58.5 + 0.96/18.02} = \frac{0.000684}{0.054} = 0.0127$$

$$x_{B2} = 1 - 0.0127 = 0.9873$$

$$M = \frac{1}{0.054} = 18.52 \text{ kg/kg mol} \qquad \frac{\rho}{M} = \frac{1027}{18.52} = 55.45 \text{ kg mol/m}^3$$

$$\left(\frac{\rho}{M}\right)_{AV} = \frac{50.05 + 55.45}{2} = 52.75 \text{ kg mol/m}^3$$

$$x_{BM} = \frac{0.9873 - 0.9114}{\ln \frac{0.9873}{0.9114}} = 0.9488$$

$$N_A = \frac{D_{AB}}{Z x_{BM}} \left(\frac{\rho}{M}\right)_{AV} (x_{A1} - x_{A2})$$

$$= \frac{1.3 \times 10^{-9}}{0.0015 \times 0.9488} \times (52.75)(0.0886 - 0.0127)$$

$$= 3.657 \times 10^{-6} \frac{\text{kg mol}}{\text{m}^2 \cdot \text{s}}$$

4.2 Assume air contains no alcohol. Therefore, $p_{A2} = 0$ bar.

The open space of 2 cm is filled with stagnant air. Thus the thickness of film is 2 cm = 0.02 m.

The partial pressure of alcohol at the interface = 0.55 bar = p_{A1}.

Therefore $p_{B1} = 0.45$ bar and $p_{B2} = 1$ bar.

$$(p_B)_{lm} = \frac{p_{B2} - p_{B1}}{\ln(p_{B2}/p_{B1})} = \frac{1.0 - 0.45}{\ln(1/0.45)} = 0.689 \text{ bar} \quad T = 273 + 10 = 283 \text{ K}$$

$$N_A = \frac{D_{AB} \times p}{RT} \times \frac{(p_{A1} - p_{A2}) \times A}{L \times (p_B)_{lm}} = \frac{0.039 \times 1 \times 0.1(0.55 - 0)}{0.08314 \times 283 \times 0.02 \times 0.689}$$

$$= 6.62 \times 10^{-3} \text{ kg mol}^2/\text{h}$$

Alcohol lost per day = $6.62 \times 10^{-3} \times 24 = 0.16$ kg mol/day

4.3
$$Re = \frac{du\rho}{\mu} = \frac{0.102(15.3)(1.2)}{0.018 \times 10^{-3}} = 104{,}040 \quad D_{AB} = 0.072 \text{ m}^2/\text{h.}$$

$$Sc = \frac{\mu}{\rho D_{AB}} = \frac{0.018 \times 10^{-3}(3600)}{1.2(0.072)} = 0.75$$

$Sh = 0.023 \; Re^{0.83} Sc^{1/3} = 0.023(104{,}040)^{0.83}(0.75)^{1/3} = 305$
$D_i = 0.102$ m

$$\text{or} \quad Sh = \frac{k_c D}{D_{AB}} = 305$$

$$\text{Therefore } k_c = \frac{0.072}{0.102} \times 305 = 215.3 \text{ m/h}$$

Vapor pressure of water at 21°C = 0.025 bar
$p_{B1} = 1 - 0.025 = 0.975$ bar $\quad p_{B2} = 1 - 0.0015 = 0.9985$ bar

$$p_{BM} = \frac{p_{B1} - p_{B2}}{\ln(p_{B1}/p_{B2})} = \frac{0.9985 - 0.975}{\ln(0.9985/0.975)} = 0.986 \text{ bar}$$

$$k_G = \frac{k_c}{RT} = \frac{215.3}{0.08314 \times 294} = 8.81 \quad \text{kg mol/(h} \cdot \text{m}^2 \cdot \text{bar} \quad \text{kg mol/h} \cdot \text{m}^2 \cdot \text{bar)}$$

$$N_A = k_G(p_{A1} - p_{A2}) = 8.81 \left[\frac{\text{kg mol}}{\text{h} \cdot \text{m}^2 \cdot \text{bar}} \right] \times (0.025 - 0.0015) \text{ (bar)}$$

$$= 0.207 \text{ kg mol/h} \cdot \text{m}^2$$

4.4 Assume Henry's law applies at 15 barA pressure. Henry's law: $p = Hc$.

Then concentration at the inner surface of pipe = $p/Hc = 15/1.67 = 8.982$ kg mol/m^3

The concentration at the outer surface of the pipe can be taken as zero.

Since hydrogen is diffusing through radially, the effective average area for diffusion must be estimated. For a cylinder, the average diffusional area is given by

$$S_{av} = \frac{2\pi L (r_2 - r_1)}{\ln(r_2/r_1)}$$

Where r_2 and r_1 are outer and inner radii respectively.

$$r_2 = 80 \times 10^{-3} \text{ m} \quad \text{and} \quad r_1 = 58.5 \times 10^{-3} \text{ m}$$

$$\text{Therefore } S_{av} = \frac{2\pi(100)(89 \times 10^{-3} - 58 \times 10^{-3})/2}{\ln(89 \times 10^{-3}/58.5 \times 10^{-3})}$$

$$= 22.835 \text{ m}^2/100 \text{ m length}$$

$$Z = \frac{1}{2}(89 - 58.5) = 15.25 \text{ mm} = 15.25 \times 10^{-3} \text{ m}$$

Molar flux (assuming diffusivity to be constant) is given by

$$N_A = \frac{0.3 \times 10^{-12} \times 22.835\,(8.982 - 0)}{15.25 \times 10^{-3}} = 4.035 \times 10^{-9} \text{ kg mol/s}$$

Mass flux = $4.035 \times 2 \times 10^{-9} = 8.07 \times 10^{-9}$ kg/s H_2 per 100 m length.

Therefore, mass flux per hour = $8.07 \times 10^{-9} \times 3600$
$$= 2.905 \times 10^{-5} \text{ kg/h per 100 m length.}$$

CHAPTER 5

Chemical Kinetics

OUTLINE

THERMODYNAMICS OF CHEMICAL REACTIONS 72

CLASSIFICATION OF CHEMICAL REACTIONS 73

CHEMICAL EQUILIBRIUM 73

RATE OF CHEMICAL REACTION (HOMOGENEOUS REACTIONS) 75
The Law of Mass Action

MOLECULARITY AND ORDER OF REACTION 76

IRREVERSIBLE REACTIONS 76

REVERSIBLE HOMOGENEOUS REACTIONS 76

COMPLEX REACTIONS 78

PARALLEL REACTIONS 78

CONSECUTIVE REACTIONS 79

HOMOGENEOUS CATALYZED REACTIONS 79

AUTOCATALYTIC REACTIONS 80

RATE EXPRESSIONS IN TERMS OF FRACTIONAL CONVERSION X 80
Irreversible Reactions ■ Reactions in Parallel ■ Reversible Reactions

MICHAELIS-MENTEN EQUATION 82

CONSTANTS OF THE RATE EQUATIONS 82
Integral Method of Analysis ■ Method of Differentiation ■ Method of Half-Times ■ Method of Reference Curves

METHOD OF k CALCULATION 83

EFFECT OF TEMPERATURE ON RATE OF REACTION 84

REACTOR DESIGN FOR HOMOGENEOUS REACTIONS 84

BATCH REACTORS 84
Flow Reactors

REFERENCES 87

Design and operation of chemical reactors to convert suitable raw materials (reactants) into useful and marketable products economically is one of the most important tasks facing a chemical engineer. To accomplish the task it is necessary to specify the type and size of the reactor system, the desired extent of reaction (approach to equilibrium or maximum conversion), and the operating conditions, such as pressure and temperature. The problem of maximum yield or equilibrium conversion and heat transfer required from or to the reactor system is adequately addressed by thermodynamics. Chemical kinetics deals with the mechanisms and rate of chemical reactions. In this chapter, we review basic principles of chemical kinetics and their application in the design of different types of reactors.

THERMODYNAMICS OF CHEMICAL REACTIONS

We begin with a review of some key thermodynamic aspects of chemical reactions. Some of these terms have been already reviewed in Chapters 2 and 3, but because of their importance in chemical kinetics, we will revisit them here.

Heat of chemical reaction is the change in enthalpy of the reaction system at constant pressure. This is a function of the nature of reactants and products involved and also their physical states.

Standard heat of reaction is the change of enthalpy when the reaction takes place at 1 atm in such a manner that it starts and ends with all involved materials in their normal states of aggregation at a constant temperature of 25 °C.

Heat of formation of a chemical compound is the change in enthalpy when elements combine to form the compound, which is the only reaction product. Standard heat of formation of a compound, H_f, is the heat of reaction at 25 °C and 1 atm.

Standard heat of combustion, H_c, is the enthalpy change resulting from the combustion of a substance in its normal state at 25 °C and atmospheric pressure, with the combustion beginning and ending at 25 °C. Generally gaseous CO_2 and liquid water are the combustion products.

Standard heat of reaction, H_R, can be calculated from the heats of formation of the reactants and the products by the relation

$$\Delta H_R = \Sigma \Delta H_{f(products)} - \Sigma \Delta H_{f(reactants)} \quad \text{at 25 °C, 1 atm}$$

It can also be calculated from the heats of combustion of the reactants and products by the relation

$$\Delta H_R = \Sigma \Delta H_{c(reactants)} - \Sigma \Delta H_{c(products)} \quad \text{at 25 °C, 1 atm}$$

Heat of reaction at any temperature can be expressed as

$$\Delta H_T = \Delta H_{T_0} + \int_{T_0}^{T} \Delta C_p dT \quad \text{where} \quad \Delta C_P = \Sigma(n_i C_{pi})_{products} - \Sigma(n_i C_{pi})_{reactants}$$

If $C_P^0 = a + \beta T + cT^2 + \cdots$ for each component taking part in the reaction, and if the reaction is of the following type:

$$n_A A + n_B B + \cdots \to n_C C + n_d D + \cdots,$$

the effect of pressure on the heat of reaction is generally negligible.

CLASSIFICATION OF CHEMICAL REACTIONS

Chemical reactions can be classified in many ways. Of more interest is the classification based on the number and types of phases involved. On this basis, reactions can be divided into two groups: (1) homogeneous reactions and (2) heterogeneous reactions. Homogeneous reactions are those that proceed in one phase. When two or more phases are present, the reaction is called heterogeneous.

Another important classification is based on the mechanism of the chemical reaction. These classifications of reactions are (1) irreversible, (2) reversible, (3) simultaneous or parallel, and (4) consecutive. Other classifications are based on order of reaction, thermal operating condition, molecularity, catalysis or no catalysis, and type of equipment used. Lastly, the classification based on mode of operation, namely, batch, semi-batch, or continuous, is also important from the point of design.

CHEMICAL EQUILIBRIUM

In reversible reactions, the reactants combine and produce products, which recombine to produce the reactants. When the forward and backward reaction rates are equal, the system reaches a state of dynamic equilibrium, and there is no net production of products. At equilibrium, maximum conversion of reactants to products is attained. This maximum conversion is called equilibrium conversion.

When the reaction is taking place at equilibrium, the temperature and pressure are constant and the change in free energy is zero. Using these conditions, a relation between the change in free energy and the equilibrium constant K is developed. This relation is given by the equation

$$G^0 = -RT \ln K$$

where K = equilibrium constant
T = absolute temperature
R = gas law constant

For a gaseous reaction of the type $aA + bB - cC + dD$, the equilibrium constant K is given in terms of the activities of the products and the reactants. The equilibrium constant is

$$K = \frac{a_C^c a_D^d}{a_A^a a_B^b}$$

where a = activities of the components

$$a_i = \frac{f_i}{f_i^0}$$

If the standard state is unit fugacity

$$f_i^0 = 1, \quad K = \frac{f_C^c f_D^d}{f_A^a f_B^b}$$

For ideal gas behavior

$$K_p = \frac{p_C^c p_D^d}{p_A^a p_B^b}$$

where p = partial pressures.
In terms of mole fractions

$$K_P = \frac{(y_c P_t)^c (y_d P_t)^d}{(y_a P_t)^a (y_b P_t)^b} = K_y P_t^{[(c+d)-(a+b)]}$$

If $P_t = 1$ atm, $K_P = K_y$

van't Hoff gives

$$\frac{d \ln K}{dT} = \frac{\Delta H_T^0}{RT^2}$$

If ΔH_T^0 is independent of temperature, integration of van't Hoff's equation gives the following relation:

$$\ln \frac{K_2}{K_1} = \frac{-\Delta H_T^0}{R} \left(\frac{1}{T_2} - \frac{1}{T_1} \right)$$

If ΔH_T^0 varies with temperature,

$$\ln K = \frac{-I_H}{RT} + \frac{\Delta a}{R} \ln T + \frac{\Delta \beta}{R} \left(\frac{1}{2} \right) T + \frac{\Delta \gamma}{R} \left(\frac{1}{6} \right) T^2 \cdots + I$$

$$\Delta G_T^0 = I_H + I_G T - \Delta a T \ln T - \frac{1}{2} \Delta \beta T^2 - \frac{1}{6} \Delta \gamma T \quad \text{where } I_G = -(IR)$$

(a) $\Delta G_T^0 < 0$, spontaneous reaction

(b) $\Delta G_T^0 = 0$, equilibrium condition

(c) $\Delta G_T^0 > 0$, no reaction

The preceding expressions permit the computation of the variation in the equilibrium constant, hence conversion, with temperature. The foregoing equations require thermodynamic data. Extensive data on free energy changes, ΔG^0 at 298 K, have been tabulated. If free energy data are not available, the thermodynamic relation $\Delta G^0 = \Delta H - T\Delta S$ can be used to establish a value for ΔG^0. In this case, specific heat data must be available as a function of temperature to compute enthalpy difference, ΔH, and entropy difference, ΔS. Entropy changes and absolute entropies are calculated using the third law of thermodynamics. Generalized property departure charts also can be used to get enthalpy and entropy changes.

RATE OF CHEMICAL REACTION (HOMOGENEOUS REACTIONS)

From the point of reactor design, there are two paramount questions that a chemical engineer needs answers to. (1) What is the equilibrium condition or what is the equilibrium conversion? (2) How quickly can a desirable conversion be attained or to what extent can the equilibrium be approached and how rapidly? The first question is answered through thermodynamics. Reaction kinetics answers the second question by finding the conditions under which a thermodynamically feasible reaction will take place at an acceptable speed. In other words, reaction kinetics deals with the rate and chemical mechanism of reaction.

Chemical reaction rate for a homogeneous (single phase) reaction is quantitatively expressed as the number of units of mass (usually moles) of some reactant or product species that is converted or produced in the reaction per unit time per unit volume of the reactor system. Mathematically the rate can be expressed by the following equation:

$$r = -\frac{1}{V}\frac{dn}{dt} \quad \text{general expression}$$

$$r = -\frac{d(n/V)}{dt} = -\frac{dC}{dt} \quad \text{at constant volume of reaction space}$$

Based on reactant, rate of reaction $= r = -\dfrac{1}{V}\dfrac{dn}{dt}$

Based on product, rate of reaction $= r = \dfrac{1}{V}\dfrac{dn}{dt}$

For general reaction, $aA + bB \rightarrow cC + dD$,

$$-\frac{1}{a}\frac{dn_A}{dt} = -\frac{1}{b}\frac{dn_B}{dt} = \frac{1}{c}\frac{dn_C}{dt} = \frac{1}{d}\frac{dn_D}{dt}$$

if ζ is extent of reaction, $d\xi = -\dfrac{dn_A}{a} = -\dfrac{dn_B}{b} = \dfrac{dn_C}{c} = \dfrac{dn_D}{d} = \dfrac{dn_i}{v_i}$

Where v_i is the stoichiometric coefficient of species i

$$r_i = \frac{v_i}{V}\frac{d\xi}{dt} \quad \text{also} \quad \frac{r_i}{v_i} = \frac{1}{v_i}\frac{dC_i}{dt}$$

Conversion X_A of component A is defined as its fraction that is converted in the reaction. Thus for the reaction $aA + bB \delta cC + dD$, the conversion X_A of component A is

$$X_A = \frac{(n_A)_0 - (n_A)}{(n_A)_0}$$

where $(n_A)_0$ = original moles of A

(n_A) = moles of A remaining unreacted

$(n_A)_0 - (n_A)$ = moles of A that have reacted

Conversion X_A is related to the extent of reaction by $X_A = \dfrac{a\xi}{(n_A)_0}$.

Rate in terms of X_A is given by $-r = \dfrac{1}{V}\dfrac{dX_A}{dt}$, X_A = fraction converted, x = moles converted.

The Law of Mass Action

The law of mass action states that the rate of a chemical reaction is proportional to the active mass of the participating reactant. Thermodynamic activity is generally considered as the active mass.

However, it should be noted here that it has not been possible so far to show experimentally that the thermodynamic and kinetic activities are the same.

MOLECULARITY AND ORDER OF REACTION

Molecularity of an elementary reaction is the number of molecules that take part in the reaction. This number is found to be either one or two in most elementary reactions, but occasionally it is found to be three.

Order of reaction with respect to a particular reactant is the power to which its concentration is raised in the rate equation. Thus if the rate equation is

$$r_A = kC_A^a C_B^b \cdots C_D^d$$

then the order of the reaction with respect to component A is a, with respect to component B is b, and so on. The overall order n of the reaction is

$$n = a + b + \cdots + d$$

Since the rate equation is empirically established, the powers of concentrations in the rate equation may not be integers only and can be fractional values. The powers $a, b, \ldots d$ are not related to stoichiometric coefficients.

IRREVERSIBLE REACTIONS

Irreversible reactions proceed in one direction only and there is no measurable reconversion of products into reactants. They are characterized by a large value of equilibrium constant. Table 5.1 summarizes reaction rate equations in terms of concentration for some irreversible homogenous reactions. Solutions of the equations by integration and the half-life periods are also listed. Half-life or half-time of a reaction denoted by $t_{1/2}$ is the time required to reduce the concentration of the reactants to one-half of the original value.

REVERSIBLE HOMOGENEOUS REACTIONS

In a reversible reaction, reactants continually combine to form products and in reverse the products combine to convert back to reactants. Equilibrium is reached when the rates of forward and reverse reactions become equal. Therefore, complete conversion of the reactants to products is not possible. In practice, attainment of complete equilibrium may take a long time. As a result, actual reactor design may

Table 5.1 Rate equations in terms of concentration (irreversible homogeneous reactions)

Reaction	Order	Rate Equation	Solution of Integral	Half-Life
$A \rightarrow P$	0	$-\dfrac{dC_A}{dt} = k_0$	$C_{A0} - C_A = k_0 t$	$t_{1/2} = \dfrac{C_{A0}}{2k_0}$
$A \rightarrow P$	1	$-\dfrac{dC_A}{dt} = k_1 C_A$	$\ln \dfrac{C_A}{C_{A0}} = -k_1 t$	$t_{1/2} = \dfrac{1}{k_1} \ln 2$
$2A \rightarrow P$	2	$-\dfrac{dC_A}{dt} = k_2 C_A^2$	$\dfrac{1}{C_A} - \dfrac{1}{(C_A)_0} = k_2 t$	$t_{1/2} = \dfrac{1}{k_2 C_{A0}}$

have to be based on some acceptable approach to equilibrium so that the design is economically feasible and time-wise practical. We will review here kinetics of some simple reversible reactions.

The reaction $A \underset{k_1}{\overset{k_2}{\rightleftarrows}} B$ is first order in both directions. The rate equation is

$$-\frac{dC_A}{dt} = \frac{dC_A}{dt} = k_1 C_A - k_2 C_B$$

The solution of this equation in terms of equilibrium concentration is given by

$$\frac{C_A - C_{Ae}}{C_{Ao} - C_{Ae}} = e^{-k_R t}$$

where $\quad k_R = \dfrac{k_1(K+1)}{K}, \quad C_{Ae} = \dfrac{C_{Bo} + C_{Ao}}{K+1}, \quad$ and $\quad K = \dfrac{k_1}{k_2}$

The reaction $A + B \underset{k_1}{\overset{k_2}{\rightleftarrows}} C + D$ is second order in both directions. The rate equation is

$$-\frac{dC_A}{dt} = k_2 \left[C_A C_B - \frac{1}{K} C_C C_D \right]$$

where K is equilibrium constant

Assuming only A and B are present initially, and choosing C_C as variable, the concentrations of A, B, and D in terms of C_C are

$$C_A = (C_A)_0 - C_C \qquad C_B = (C_B)_0 - C_C \qquad C_D = C_C$$

With these substitutions, the rate equation becomes

$$\frac{1}{k_2} \frac{dC_C}{dt} = [(C_A)_0 - C_C][(C_B)_0 - C_C] - \frac{1}{K} C_C^2$$

This is transformed to

$$\frac{1}{k_2}\frac{dC_C}{dt} = \alpha + \beta C_C + \gamma C_C^2$$

where, $\alpha = (C_A)_0(C_B)_0$,

$\beta = -[(C_A)_0 + (C_B)_0]$

$\gamma = 1 - 1/K$

Solution with boundary condition $C_C = 0$ at $t = 0$, is

$$\frac{1}{q^{1/2}}\ln\frac{2\gamma C_C/(\beta - q^{1/2}) + 1}{2\gamma C_C/(\beta + q^{1/2}) + 1} = k_2 t$$

where $q = \beta^2 - 4\alpha\gamma$

COMPLEX REACTIONS

Reactions are considered complex when stable products are produced by more than one reaction. Parallel or simultaneous and consecutive reactions are examples of complex reactions. Yield of a specific product is defined as the fraction of the reactant that produces that specific product. Two other concepts, point selectivity and overall selectivity, are defined in the following manner

$$\text{Point selectivity} = \frac{\text{Rate of production of one product}}{\text{Rate of production of another product}}$$

$$\text{Overall selectivity} = \frac{\text{Amount of one product produced}}{\text{Amount of another product produced}}$$

PARALLEL REACTIONS

A simple case of parallel reactions occurs when a single component, A, converts to give two products B and C. Rate equations for these reactions are

$$A \xrightarrow{k_1} B \qquad A \xrightarrow{k_2} C$$

(1) $-\dfrac{dC_A}{dt} = (k_1 + k_2)C_A$ (2) $\dfrac{dC_B}{dt} = k_1 C_A$ and (3) $\dfrac{dC_C}{dt} = k_2 C_A$

With the condition that at $t = 0$, $C_A = C_{A0}$ and $C_B = C_C = 0$,

$$x_B \frac{C_B}{C_{A0}} = \frac{k_1}{k_1 + k_2} = \left(1 - \frac{C_A}{C_{A0}}\right) = \frac{k_1}{k_1 + k_2} x_t$$

and $$x_C = \frac{C_C}{C_{A0}} = \frac{k_2}{k_1 + k_2}\left(1 - \frac{C_A}{C_{A0}}\right) = \frac{k_2}{k_1 + k_2} x_t$$

Where x_t is the total conversion of A to both B and C

The selectivity of B with respect to C is $S_B = \dfrac{x_B}{x_C} = \dfrac{k_1}{k_2}$

CONSECUTIVE REACTIONS

In consecutive reactions, a reactant converts to a product, which in turn converts to another product. As an example, in the following set of reactions, A converts to B, which reacts to produce C. Such reactions are also called reactions in series.

$$A \xrightarrow{k_1} B \xrightarrow{k_2} C$$

The rate equations for these can be written down as

$$(1)\ \frac{dC_A}{dt} = -k_1 C_A \quad (2)\ \frac{dC_B}{dt} = k_1 C_A - k_2 C_B \quad (3)\ \frac{dC_D}{dt} = k_2 C_B$$

With initial condition: at $t = 0$, $C_A = C_{A0}$, $C_B = C_C = 0$, solution of these equations is

$$C_A = C_{A0} e^{-k_1 t}, \quad C_B = C_{A0} k_1 \left\{ \frac{e^{-k_1 t}}{k_2 - k_1} - \frac{e^{-k_2 t}}{k_1 - k_2} \right\}$$

and $\quad C_D = C_{A0} \left\{ 1 + \dfrac{k_2 e^{-k_1 t}}{k_1 - k_2} + \dfrac{k_1 e^{-k_2 t}}{k_1 - k_2} \right\}$

Maximum concentration of B occurs at

$$t_{max} = \frac{\ln(k_2/k_1)}{k_2 - k_1} \quad \text{and} \quad \frac{C_{B\,max}}{C_{A0}} = \left(\frac{k_1}{k_2} \right)^{k_2/(k_2 - k_1)}$$

HOMOGENEOUS CATALYZED REACTIONS

In catalyzed reactions, it is assumed that the rate of overall reaction is the sum of the rates of the uncatalyzed and catalyzed reactions.

Reaction: $A \xrightarrow{k_1} PA + C \xrightarrow{k_2} P + C$ where C = catalyst

Reaction rates $\quad (1)\ -\left(\dfrac{dC_A}{dt}\right)_1 = k_1 C_A \quad (2)\ -\left(\dfrac{dC_A}{dt}\right)_2 = k_2 C_A C_C$

The overall rate of disappearance of reactant A is then

$$-\frac{dC_A}{dt} = k_1 C_A + k_2 C_A C_C = (k_1 + k_2 C_C) C_A$$

Since catalyst concentration remains constant, the solution in terms of conversion is

$$-\ln \frac{C_A}{C_{A0}} = -\ln(1 - X_A) = (k_1 + k_2 C_C) t$$

AUTOCATALYTIC REACTIONS

In autocatalytic reactions, one of the products of the reaction acts as a catalyst. The simplest reaction of this type is given next.

Reaction $\quad A + R \to R + R \quad$ R is product as well as catalyst

Rate equation is $\quad \dfrac{dC_A}{dt} = kC_A C_R$

At any time, $C_0 = C_A + C_R = C_{A0} + C_{R0} =$ constant.
Therefore the rate equation becomes

$$-r_A = -\dfrac{dC_A}{dt} = kC_A(C_0 - C_A)$$

After rearrangement, breaking into partial fractions and integration gives the solution

$$\ln \dfrac{C_{A0}(C_0 - C_A)}{C_A(C_0 - C_{A0})} = \ln \dfrac{C_R/C_{R0}}{C_A/C_{A0}} = C_0 kt = (C_{A0} + C_{R0})kt$$

RATE EXPRESSIONS IN TERMS OF FRACTIONAL CONVERSION X

Fractional conversion was defined in an earlier section. In many situations, it is used as a more convenient variable to work with. Therefore, the rate equations and their solutions in terms of fractional conversion are presented for some simple reactions of all types as is done with concentration as the variable.

Irreversible Reactions

A few cases of irreversible reactions are considered in this subsection. Rate equations and solutions to the equations are given in terms of X.

First order reaction $A \; \tau \; P$

Rate expresison $\quad \dfrac{dX_A}{dt} = k(1 - X_A)$

Solution: $\quad -\ln(1 - X_A) = kt \quad$ at $t = 0$, $X_A = 0$

Second Order Reaction: $A + B \to P$

Rate expression: $\quad C_{Ao}\dfrac{dX_A}{dt} = kC_{Ao}^2(1 - X_A)(M - X_A) \quad M = C_{Bo}/C_{Ao}$

when $M \neq 1$, solution is $\quad \ln \dfrac{M - X_A}{M(1 - X_A)} = C_{Ao}(M - 1)kt$

when $M = 1$, rate expression $\quad \dfrac{dX_A}{dt} = kC_{Ao}(1 - X_A)^2$

Solution with boundary condition, $t = 0$, $X_A = 0$, $\quad \dfrac{1}{C_{Ao}} \dfrac{X_A}{1 - X_A} = kt$

n^{th} order:

$$\text{Rate expression:} \quad C_{A0} \dfrac{dX_A}{dt} = kC_{A0}^n (1 - X_A)^n$$

B.C. $t = 0$, $X_A = 0$; Solution: $(1 - X_A)^{1-n} - 1 = (n-1)C_{A0}^{n-1} kt$

Zero order:

$$\text{Rate expression} \quad \dfrac{dX_A}{dt} = k$$

B.C. $t = 0$, $X_A = 0$; Solution is: $C_{A0} X_A = kt$

Reactions in Parallel

One simple case of two reactions in parallel is considered here.

$$\text{The reactions:} \quad A \xrightarrow{k_1} B \qquad A \xrightarrow{k_2} C$$

B.C. at $t = 0$, $C_A = C_{A0}$, $C_B = C_{B0}$ and $C_C = C_{C0}$

Solution in terms of conversion: $-\ln(1 - X_A) = (k_1 + k_2) t$

Reversible Reactions

In reversible reactions, the conversion, X, is expressed in terms of the maximum or equilibrium conversion, X_e. We consider here two simple cases of reversible reactions.

First order reversible:

$$A \underset{k_1}{\overset{k_2}{\rightleftharpoons}} B$$

$$\text{Rate expression:} \quad \dfrac{dX_A}{dt} = \dfrac{k_1(M+1)}{M + X_{Ae}}(X_{Ae} - X_A) \qquad M = C_{B0}/C_{A0}$$

$$\text{Solution:} \quad -\ln\left(1 - \dfrac{X_A}{X_{Ae}}\right) = \dfrac{M+1}{M + X_{Ae}} k_1 t$$

where X_{Ae} is equilibrium conversion of component A.

Second order reversible:

$$A \underset{k_1}{\overset{k_2}{\rightleftharpoons}} B$$

$$\text{Rate equation:} \quad \dfrac{dX_A}{dt} = k_1 C_{Ao}^2 (1 - X_A)^2 - \dfrac{k_2}{k_1} X_A^2$$

$$\text{Solution:} \quad \ln \dfrac{X_{Ae} - (2X_{Ae} - 1)X_A}{X_{Ae} - X_A} = 2 k_1 C_{Ao} \left(\dfrac{1}{X_{Ae}} - 1\right) t$$

MICHAELIS-MENTEN EQUATION

Sometimes experimental data correlate very well by one reaction order at high reactant concentrations and by another at low concentrations. These are reactions of *shifting order*. Certain forms of the rate equation fit this type of data. One such equation developed by Michaelis-Menten while studying enzyme-catalyzed reactions is

$$-r_A = -\frac{dC_A}{dt} = \frac{k_1 C_A}{1 + k_2 C_A}$$

for a reaction $A \rightarrow R$.

At high concentrations ($k_2 C_A \gg 1$), the reaction is of zero order with rate constant k_1/k_2. At low concentrations ($k_2 C_A \ll 1$), the reaction is of the first order with rate constant k_1. Solution of the equation is

$$\frac{\ln(C_{A0}/C_A)}{C_{A0} - C_A} = -k_2 + \frac{k_1 t}{C_{A0} - C_A}$$

Two more general forms of the preceding equation are

$$-r_A = -\frac{dC_A}{dt} = \frac{k_1 C_A^m}{1 + k_2 C_A^n} \quad \text{and} \quad -r_A = -\frac{dC_A}{dt} = \frac{k_1 C_A^m}{(1 + k_2 C_A)^n}$$

These forms can also be used to fit experimental data of two orders. A study of mechanism will show the form to use. These equations are used to represent the kinetics of surface-catalyzed reactions.

In some reactions, the shift occurs from high to low order as the concentration drops. This behavior can be explained by considering two competing reaction paths of two different reaction orders. For example, for *zero and first order* decompositions of the reaction $A \rightarrow R$, the overall disappearance of A is the sum of the individual rates or

$$(-r_A)_{overall} = -\left(\frac{dC_A}{dt}\right)_{overall} = k_1 + k_2 C_A$$

The integrated form of this equation is

$$-\ln\left(\frac{k_1 + k_2 C_{A0}}{k_1 + k_2 C_A}\right) = k_2 t$$

Other reaction orders can also be treated by this type of analysis.

CONSTANTS OF THE RATE EQUATIONS

So far we have reviewed mainly the mechanism of a reaction and how it can be used to establish the rate equation. Now we review the methods employed to determine the values of the constants in the rate equations. Experimental chemical kinetic data are usually obtained in terms of concentrations at constant temperature and volume as a function of time. To use the rate equations in the design of reactors, it is necessary to determine the rate constants from experimental data.

In most cases, the stoichiometry of the reaction suggests the form of the equation to use first. When the mathematical equation is written, the next step is to find the constants in the equation. A number of methods are used to accomplish this objective.

Integral Method of Analysis

In the integral method of analysis, a particular form of the rate equation is assumed based on the stoichiometry of the reaction. Its solution is obtained by integration with appropriate boundary conditions. By examining the integration equation, prediction of the form of concentration function is made that when plotted against time would yield a straight line. From this plot, the rate constants are then determined.

Method of Differentiation

In this method, the rate data are tested by directly fitting the rate expression to the data. To accomplish this, the rates of reaction are calculated from the experimental data by graphical or numerical means and plotted against appropriate function of concentration indicated by the unintegrated equation. A straight line indicates a good fit. From this plot, the rate constants are determined.

Method of Half-Times

At 50% conversion, the integrated equations assume simple forms. Since half-time is a function of initial concentration, data must be collected with several initial concentrations to evaluate both k and n (order of reaction). A plot of log $t_{1/2}$ versus log C_{A0} yields a straight line.

Method of Reference Curves

This method is particularly useful to determine order of a reaction. Noddings prepared the reference curves required for this method, which is a plot of percent conversion versus $t/t_{0.9}$ with n (order of reaction) as parameter, where t is time for the particular conversion and $t_{0.9}$ is the time required for 90% conversion. To use this method the experimental data are plotted in the same way as on the reference plot and on the same scale. Superimposition of the actual plot on the reference plot suggests the order of the reaction.

METHOD OF k CALCULATION

In this method a value for the order of the reaction is assumed and k values are calculated at various experimental data points. If calculated k values are nearly constant, the assumed order of reaction n is correct.

EFFECT OF TEMPERATURE ON RATE OF REACTION

In the preceding sections, the effect of concentration of reactants and products on the rate of reaction was considered at constant temperature and volume. In this section, we consider the effect of temperature on reaction rate.

Two important relations describe the effect of temperature on the rate of reaction for elementary reactions.

1. Arrhenius relation: $k = Ae^{-E/RT}$
   where A is the frequency factor
   E is the activation energy

2. From transition-state or collision theory: $k \propto T^m e^{-E/RT}$

where $m = \tfrac{1}{2}$ in case of collision theory and $m = 1$ in case of transition-state theory. For nonelementary reactions, the composite rate constants also vary as $e^{-E/RT}$. A plot of $\ln k$ vs. $1/T$ gives a straight line with slope of $-E/R$ from which the activation energy E can be computed.

The temperature dependence of the equilibrium constant of the elementary reversible reactions is given by the van't Hoff equation

$$\frac{d(\ln K)}{dT} = \frac{\Delta H_R}{RT^2}$$

REACTOR DESIGN FOR HOMOGENEOUS REACTIONS

Homogeneous reactions are usually carried out in batch, continuous, or semi-batch reactors. For the design of a reactor system, we need to know first the type of the reactor to use and the best method of its operation. The next step includes calculation of reactor size, composition of product stream, and specification of the required operating conditions, such as temperature, pressure, and compositions. Final design will be based on profitability of the project, safety, and environmental impact. The data usually available are the temperature, pressure, flow rates of feeds, and the required production rate.

A reactor design starts with a *mass balance* for any reactant or product. *The applicable rate equation reviewed in foregoing sections enters the mass balance equation.* If the composition within the reactor is uniform, mass balance can be written over the whole reactor. If it is not uniform, the mass balance will have to be done on a differential element of the reactor and integrated over the whole reactor for the given flow and concentration conditions. For various types of reactors, the mass balance equation simplifies and on integration yields the basic *performance equation* for that particular unit type. If the operation is nonisothermal, an *energy balance* is also necessary.

In the following section, we briefly review the *performance equations* for various reactor types.

BATCH REACTORS

In a batch operation, the reactants are charged into the reactor, thoroughly mixed, and left to react for a certain period after which the reaction mixture is discharged. The batch operation is essentially an unsteady operation. The composition in the reactor changes with time but at any given time it is uniform throughout the reactor.

Material balance for a component A results in

$$(-r_A)V = N_{AO}\frac{dX_A}{d\theta} \quad \text{where } \theta = \text{time}$$

Solution: $\theta = N_{AO}\int_0^{X_A}\frac{dX_A}{(-r_A)V}$

When density is constant, $\theta = C_{AO}\int_0^{X_A}\frac{dX_A}{-r_A} = -\int_{C_{AO}}^{C_A}\frac{dC_A}{-r_A}$

For reactions in which volume changes linearly with conversion

$$\theta = C_{AO}\int_0^{X_A}\frac{dX_A}{(-r_A)(1+\varepsilon_A X_A)}$$

where ε_A = fractional volume change between no conversion and complete conversion

Flow Reactors

Basically flow reactors are of two types: (1) steady state mixed flow reactors or continuous flow stirred tank reactors (CFSTRs) and (2) steady state tubular or plug flow reactors. To meet the required production rate, they may be single units or a combination of individual units in parallel or in series because of the practical constraints imposed on the size or length of a single reactor vessel or tube. In flow reactors, the reactants are fed continuously to the reactor and products mixed with un-reacted feeds are also withdrawn continuously.

For flow reactors, *space-time* and *space velocity* are *performance measures*. They are defined as follows:

Space-time:

$\tau = \dfrac{1}{s}$ = Time required to process one reactor volume of feed measured at specified conditions = (time)

Space velocity: The space velocity is related to space-time by

$S = \dfrac{1}{\tau}$ = the number of reactor volumes of feed at specified conditions that can be treated in unit time = $(\text{time})^{-1}$

For the conditions of the feed, $\tau = \dfrac{1}{S} = \dfrac{C_{AO}V}{F_{AO}}$

$$= \frac{(\text{moles } A \text{ feed/volume of feed})(\text{volume of reactor})}{\text{moles } A \text{ feed/time}}$$

$$= \frac{V}{v_o} = \frac{\text{reactor volume}}{\text{volumetric feed rate of } A}$$

Steady-state mixed flow reactors or continuous-flow stirred tank reactors (CFSTR)

This reactor is also called a backmix reactor. It is a reactor in which contents are stirred and continuously mixed. The concentration of the exit stream from this reactor is the same as that of the reactor contents.

Steady state material balance gives the result

$$\tau = \frac{1}{S} = \frac{V}{v_o} = \frac{VC_{AO}}{F_{AO}} = \frac{C_{AO}X_{AO}}{-r_A} \quad X_A = 0 \text{ in feed.}$$

If feed is partially converted,

$$X_A \neq 0, \quad \frac{V}{F_{AO}} = \frac{X_{Af} - X_{Ai}}{(-r_A)_f}$$

Where f and i denote the exit and inlet conditions
For the special case where the density is constant

$$\tau = \frac{V}{v} = \frac{C_{AO} - C_A}{-r_A}$$

Steady-state tubular (plug-flow) reactors

In an ideal tubular flow reactor, there is no mixing in the direction of flow and complete mixing perpendicular to the direction of flow. Tubular flow reactors are usually operated at steady state so that properties are constant with respect to time. A steady state material balance yields the relation

$$V = F_{AO} \int_0^{X_A} \frac{dX_A}{-r_A}$$

which gives plug-flow reactor volume for conversion X.

$$\text{Then} \quad \tau = \frac{V}{v_o} = C_{AO} \int_0^{X_A} \frac{dX_A}{-r_A}$$

If feed is partially converted,

$$\frac{V}{F_{AO}} = \frac{V}{C_{AO}v_o} = \int_{X_{Ai}}^{X_{Af}} \frac{dX_A}{-r_A} \quad \text{or} \quad \tau = \frac{V}{v_o} = C_{AO} \int_{X_{Ai}}^{X_{Af}} \frac{dX_A}{-r_A}$$

For the special case when density is constant

$$\tau = \frac{V}{v_o} = -\int_{C_{AO}}^{C_{Af}} \frac{dC_A}{-r_A}$$

Mixed-flow reactors in series

This is a battery of CFSTRs arranged in series where the reactants are fed continuously to the first tank, and the overflow from one tank is fed to the next also continuously in succession. The product is withdrawn from the last tank in the series. Thorough mixing is maintained in each tank so that in each tank the composition is uniform. Sometimes, several stages of reactors are accommodated in a single vessel, vertical or horizontal.

For a first order reaction system and equal volume reactors in series, with assumption of constant density, a material balance on reactor i for component A gives

$$\tau = \frac{C_o V}{F_o} = \frac{V}{v} = \frac{C_o(X_i - X_{i-1})}{-r_A} \quad \text{or} \quad \tau = \frac{C_{Ai-1} - C_{Ai}}{K C_{Ai}}$$

Since space-time is the *same for all reactors*,

$$\frac{C_{Ai-1}}{C_{An}} = (1+k\tau)^n \quad \text{or} \quad C_{An} = C_{AO}(1+k\tau)^{-n}$$

For the reactor system as a whole

$$\tau_n = \frac{n}{k}\left[\left(\frac{C_{AO}}{C_{An}}\right)^{1/n} - 1\right]$$

REFERENCES

1. AIChE Modular Instruction Series, *Kinetics*, vol 1–5.
2. Das, D. K., and R. K. Prabhudesai, *Chemical Engineering License Review*, 2nd ed., Kaplan AEC Education, 2004.
3. Levenspiel, O., *Chemical Reaction Engineering*, 3rd ed., New York, John Wiley & Sons, 1998.
4. Smith, J. M., *Chemical Engineering Kinetics*, 3rd ed., New York, McGraw-Hill, 1981.

PROBLEMS

5.1 What is the heat of reaction for the following? Given are heats of combustion.

$$C_2H_5OH(l) + O_2(g) \rightarrow CH_3COOH(l) + H_2O(l)$$

$C_2H_5OH \quad \Delta H_C = -1366913$ kJ/kg mol at 25 °C

$CH_3COOH \quad \Delta H_C = -871695$ kJ/kg mol at 25 °C

5.2 Ethyl benzene decomposes according to the reaction $C_6H_5C_2H_5 \rightarrow C_6H_5C_2H_3 + H_2$.
The reaction rate constants at two temperatures are as follows:

Temp °C	k × 10⁴
540	1.6
550	2.8

Calculate the activation energy of the reaction.

5.3 Calculate free energy change for the following reaction:

$$C_2H_5OH(l) + O_2(g) \rightarrow CH_3COOH(l) + H_2O(l)$$

$(\Delta G_f^0)_{25}$ (kJ/kg mol) $-174724 \qquad -369322 \qquad -237191$

5.4 For the reaction A → B, $\Delta H_R = -100416$ kJ/kg mol and $\Delta G_b^0 = -12552$ kJ/kg mol at 25 °C. Calculate the equilibrium constant and equilibrium conversion for the reaction at 60 °C. Assume H_R is independent of temperature.

5.5 A certain reaction has the rate given by $-r_A = 0.01\ C_A^2$ gmol/cm³·min. If the rate is to be expressed in kg mol/liter²·h, what is the value and units of the rate constant?

5.6 For a gas phase elementary reaction, $A \rightarrow 3R$, what is the fractional volume change assuming the volume varies linearly between zero and complete conversion? The reaction mixture initially contains 40% by volume inserts.

5.7 A substance A decomposes by first order kinetics. In a batch reactor, 50% A is converted in 5 min. How much longer would it take to reach 90% conversion?

5.8 The first order reversible reaction $A \rightarrow R$ takes place in a batch reactor. Initial conditions are $C_{RO} = 0$, $C_{AO} = 0.75$ gmol/liter. After 10 minutes, the conversion of A is 40% and the equilibrium conversion is 65%. Develop the rate equation for this reaction.

5.9 The reaction $CO_2 + H_2 = CO + H_2O$ is carried out by heating to 1000 K and allowing the reaction to come to equilibrium at 50 bar total pressure. 60% of CO_2 is found to be converted. What is the value of the equilibrium constant K_P, and what is the partial pressure of CO in the mixture if the initial mixture consisted of only CO_2 and H_2 in equimolar proportion? Assume ideal behavior of components.

5.10 A liquid phase reaction $A \rightarrow R$ is carried out in a CFSTR. The reaction rate is 1 mol/liter/h. Feed is 5% converted. (1) Find the size of the reactor needed for 80% conversion if feed is to be processed at a rate of 1200 gmol/h. (2) If the feed rate is doubled, what is the size of the reactor needed?

5.11 A homogeneous liquid phase second-order reaction $2A \rightarrow R$ is carried out in a plug-flow reactor with 60% conversion. What will be the conversion in a plug-flow reactor two times as large if all other variables remain the same, and if the reaction takes place without a volume change? Feed to the reactor is pure A.

Chapter 5 Chemical Kinetics

SOLUTIONS

5.1
$$\Delta H_R = (\sum \Delta H_c)_{reactants} - (\sum \Delta H_C)_{products}$$
$$= -1366913 - (-871695) = -495218 \text{ kJ/kmol}$$

5.2 Arrhenius relation $\quad k = Ae^{-E/RT}$

$$T_2 = 550 + 273 = 823 \text{ K} \quad T_1 = 540 + 273 = 813 \text{ K}$$

$$\frac{k_2}{k_1} = \frac{e^{-E/RT_2}}{e^{-E/RT_1}}$$

By taking logarithms $\quad \ln\dfrac{k_2}{k_1} = -\dfrac{E}{RT_2} + \dfrac{E}{RT_1}$

From which,

$$E = \ln(k_2/k_1)\left[R\left(\frac{T_2 T_1}{T_2 - T_1}\right)\right]$$

$$= \ln\left(\frac{2.8\times 10^{-4}}{1.6\times 10^{-4}}\right)\left[8.314\times \frac{823\times 813}{823 - 813}\right] = 311308 \text{ kJ/kmol}.$$

5.3
$$\Delta G_f^0 = -369322 - 237191 + 174724 = -431789 \text{ kJ/kmol}$$

5.4 ΔH_R is independent of temperature,

Therefore, $\ln\dfrac{K_2}{K_1} = -\dfrac{\Delta H_R}{R}\left[\dfrac{1}{T_2} - \dfrac{1}{T_1}\right]$

Also, $\quad \Delta G_f^0 = -RT\ln K \quad\quad \ln K = -\Delta G_f^0/RT$

$$K = e^{-\Delta G_f^0/RT} = e^{-(-12552)/(8.314\times 298)} = 158.6 \quad T_2 = 60 + 273 = 333 \text{ K}$$

$$\ln\frac{K_2}{158.6} = \frac{-100416}{8.314}\left[\frac{1}{333} - \frac{1}{298}\right] = -4.26$$

Therefore, $K_2/158.63 = e^{-4.26} = 0.014122$

and then $K_2 = 0.014122 \times 158.62 = 2.24$

At equilibrium, $K = \dfrac{X_{AE}}{1 - X_{AE}} = 2.24$

Therefore, $X_{AE} = K/(K+1) = 0.6915$

5.5
$$-r_A = 0.01\, C_A^2 \frac{\text{gmol}}{\text{cm}^3 \cdot \text{min}}$$

$$= 0.01\, C_A^2 \frac{\dfrac{\text{gmol}}{\text{cm}^3} \cdot \dfrac{\text{kg mol}}{1000\ \text{gmol}}}{\dfrac{\text{liter}}{1000\ \text{cm}^3} \cdot \dfrac{\text{min}}{60\ \text{min}} \cdot \text{h}}$$

$$= 0.01 \times 60\, \frac{\text{kg mol}}{\text{liter} \cdot \text{h}} = 0.6\, C_A^2\ \text{kg mol/(liter} \cdot \text{h)}$$

The value of rate constant is 0.6

Since $k = -r_A/C_A^2$, the units of k are now $k = \dfrac{\text{kg mol/(liter} \cdot \text{h)}}{(\text{kg mol/liter})^2} = \dfrac{\text{liter}}{\text{kg mol} \cdot \text{h}}$

5.6 $A \rightarrow 3R$

	Initial	Final
A	0.6	0.0
R	0.0	1.8
Inerts	0.4	0.4
Total	1.0	2.2

Therefore, $\epsilon_A = \dfrac{2.2 - 1.0}{1.0} = 1.2$

5.7   In terms of conversion, $-\ln(1 - X_A) = kt$ solution of first order rate equation.

At $t = 5$ min, $X_A = 0.5$ Therefore, $-\ln(1 - 0.5) = k \times 5$
Then $k = -\ln 0.5/5 = 0.13863\ \text{min}^{-1}$

For 90% conversion, $-\ln(1 - 0.9) = 0.13863 \times t$
$t = -\ln(0.1)/0.13683 = 16.6$ min

Therefore, extra time required to reach 90% conversion $= 16.6 - 5.0 = 11.6$ min.

5.8   Rate equation $\dfrac{dX_A}{dt} = \dfrac{k_1(M+1)}{M + X_{Ae}}(X_{Ae} - X_A)$

In terms of conversion, $-\ln\left(1 - \dfrac{X_A}{X_{Ae}}\right) = \dfrac{M+1}{M + X_{Ae}} k_1 t$

In this case, $M = C_{Ro}/C_{Ao} = 0$

Therefore, $-\ln(1 - X_A/X_{ae}) = (1/X_{Ae})k_1 t$

$X_{Ae} = 0.65$, and $X_A = 0.4$ when $t = 10$ min.

$$-\ln\left(1 - \frac{0.4}{0.65}\right) = (1/0.65) \times k_1 \times (10) \quad \text{from which} \quad k_1 = 0.0621 \text{ min}^{-1}$$

Rate equation is $\dfrac{dX_A}{dt} = \dfrac{k_1(M+1)}{M + X_{Ae}}(X_{Ae} - X_A) = \dfrac{0.0621(0+1)}{(0+0.65)}[0.65 - X_A]$

Therefore, rate equation is $\dfrac{dX_A}{dt} = 0.0621 - 0.09554 X_A$

5.9 $\quad K_P = \dfrac{y_{CO} y_{H_2O}}{y_{CO_2} y_{H_2}} \times \dfrac{50 \times 50}{50 \times 50} = K_y$

	Initial	At Equilibrium
CO_2	1.0	0.4
H_2	1.0	0.4
CO	0.0	0.6
H_2O	0.0	0.6
	2.0	2.0

$y_{CO} = \dfrac{0.6}{2} = 0.3 \qquad y_{H_2O} = 0.3 \qquad y_{CO_2} = 0.4/2 = 0.2 \qquad y_{H_2} = 0.2$

$K_P = \dfrac{0.3 \times 0.3}{0.2 \times 0.2} = 2.25 \quad$ and the partial pressure of CO $= 0.3(50) = 15$ barA.

5.10 (1) For a CFSTR, $\quad \dfrac{V}{F_{Ao}} = \dfrac{X_{Af} - X_{Ai}}{(-r_A)_f}$

$V = \dfrac{X_{Af} - X_{Ai}}{(-r_A)_f} \times F_{Ao} = \dfrac{0.8 - 0.05}{1} \times 1200 = 900$ liters

(2) $\int V = 0.75(1200 \times 2) = 1800$ liters

5.11 For no volume change, $\varepsilon = 0$, $\quad -r_A = -\dfrac{dC_A}{dt} = C_{AO}\dfrac{dX_A}{dt} = kC_{AO}^2(1 - X_A)^2$

For a plug flow reactor

$$V = F_{AO}\int_0^{X_A} \dfrac{dX_A}{-r_A} = F_{AO}\int_0^{X_A}\dfrac{dX_A}{kC_{AO}^2(1-X_A)^2} = \dfrac{F_{AO}}{kC_{AO}^2}\left[\dfrac{1}{1-X_A}\right]_0^{X_A}$$

$$= \dfrac{F_{AO}}{kC_{AO}^2}\dfrac{X_A}{1 - X_A}$$

All conditions except the volume are given to be the same, therefore by comparison of the two reactor volumes, the following can be written

$$\frac{V_2}{V_1} = \frac{[X_{A'}(1-X_A)]_2}{[X_{A'}(1-X_A)]_1}$$

or

$$2 = \frac{[X_{A'}(1-X_A)]_2}{[0.6'(1-0.6)]_1}$$

Then

$$\left[\frac{X_A}{1-X_A}\right]_2 = 2 \times \left[\frac{0.6}{1-0.6}\right]_1 = 3$$

Solving,

$[X_A]_2 = 0.75$

Therefore, by doubling the reactor volume but keeping the same other conditions, 75% conversion will be obtained.

CHAPTER 6

Process Design and Economic Evaluation

OUTLINE

DEGREES OF FREEDOM FOR OPTIMIZATION 95

FIXED INVESTMENT 96

GROSS PROFIT AND NET PROFIT 96

RETURN ON INVESTMENT 96

PAYOUT OR PAYBACK TIME 97

VENTURE PROFIT AND VENTURE COST 97

LINEAR BREAK-EVEN ANALYSIS 97

ORDER-OF-MAGNITUDE ESTIMATE OF EQUIPMENT AND INSTALLED COSTS 97

The best equipment design is the least expensive and safest design that meets the specification and performs the duty. Since all businesses are motivated by profit, economic evaluation plays an important role in process design. There are many ways to reach the same destination. The best way is the least expensive way in terms of time, money, and safety.

DEGREES OF FREEDOM FOR OPTIMIZATION

Degrees of freedom for optimization (F) refers to the number of variables that are not fixed or specified in a design and therefore are free to be adjusted for optimum design.

$$F = M - N$$

where
- M = number of variables, such as equipment type, pressure, temperature, etc.
- N = number of independent sources of information, such as heat and material balances

96 Chapter 6 Process Design and Economic Evaluation

F has two components:

$$F = F_1 + F_2$$

where
- F_1 = environmental degrees of freedom. These are variables whose values are fixed by the environment of the process, such as cooling water supply temperature or the inlet and outlet temperatures of the process side of a heat exchanger, etc. This is a freedom lost by the optimizer.
- F_2 = economic degrees of freedom. These are variables that the designer can adjust to maximize the profitability, such as the type of an exchanger or the cooling water flow rate.

FIXED INVESTMENT

Fixed investment, I_F, is the investment associated with all processing equipment and auxiliary facilities that are needed for the complete operation.

GROSS PROFIT AND NET PROFIT

Gross profit, R, is expressed as follows:

$$R = S - C$$

where
- R = gross profit, $/year
- S = sales revenue, $/year
- C = cost of production, $/year

The equation for net profit, P_N, is

$$P_N = (R - dI_F)(1 - t)$$

where
- d = depreciation, decimal
- t = tax rate, decimal

RETURN ON INVESTMENT

Return on investment (ROI) is generally defined by:

$$\text{ROI} = \frac{\text{Average annual after} - \text{tax profit during earning life}}{\text{Original fixed investment} + \text{working capital}}$$

If working capital is ignored, ROI may be expressed by

$$\text{ROI} = \frac{(R - dI_F)(1 - t)}{I_F}$$

Very often, an incremental investment is necessary to reduce annual operating cost. When the sales revenue is the same, the return on such incremental investment is justified by

$$\text{ROI} = \frac{[(C + dI_F)_{\text{basecase}} - (C + dI_F)_{\text{othercase}}](1 - t)}{(I_F)_{\text{othercase}} - (I_F)_{\text{basecase}}}$$

PAYOUT OR PAYBACK TIME

Payout or payback time, T, is expressed as:

$$T = \frac{I_F}{R-(R-dI_F)t}$$

Often, for simplicity, taxes are not considered, and $T = I_F/R$

VENTURE PROFIT AND VENTURE COST

The equation for venture profit, V, is

$$V = (R-dI_F)(1-t) - i_m(I_F + I_W)$$

Venture cost, V_C, is expressed as

$$V_c = (C+dI_F)(1-t) + i_m(I_F + I_w)$$

where
- i_m = minimum acceptable rate of return, decimal
- I_w = working capital; money invested in raw materials, etc., and cash required to run the project. This may be ignored for simplified analysis.

Optimization studies generally involve maximization of venture profit and minimization of venture cost or total annual cost. Total annual cost includes fixed annual cost and variable annual cost. The fixed annual cost is independent of production and includes such items as depreciation on fixed investment and minimum allowable return on investment after income taxes, taxes and insurance, cost of safety, administration, general maintenance, and general services. The variable annual cost includes cost of raw materials, direct labor, utilities, and maintenance cost dependent on operation, warehouse, and shipping.

LINEAR BREAK-EVEN ANALYSIS

$$Q_B = \frac{F}{s-c}$$

where
- Q_B = break-even capacity, unit/time
- F = fixed cost, $/time
- s = selling price, $/unit
- c = variable cost, $/unit

ORDER-OF-MAGNITUDE ESTIMATE OF EQUIPMENT AND INSTALLED COSTS

The cost of equipment may be expressed in the form:

$$C = K(\text{Size})^n$$

where
- C = equipment cost
- K and n are constants related to the equipment type

Table 6.1 Factored Cost Exponents and Installation Cost Factors

Equipment	Size Basis	Value of Exponent, n	Lang Factor, K
Pump	HP	0.52	4.0
Shell & tube Heat exchanger	Area	0.62	3.3(ss), 4.0(cs)
Air cooler	Area	0.75	2.2(ss), 2.3(cs)
Column	Diameter	0.65	4.0
Crystallizer	Tons/day dry solid	0.55	2.4
Plate & frame filter	Area	0.58	2.7
Rotary vacuum filter	Effective area	0.63	2.4
Rotary kiln dryer	Area	0.8	3.0
Blower & fan	Cfm	0.68	2.8
Bowl centrifuge	HP	0.73	2.4
Boiler	Lbs/h	0.7	1.8
Ball mills	Tons/h	0.7	1.8
Pulverizer	Tons/h	0.39	2.5
Atm. storage tank	Gallon	0.6	4.0
Pressure vessel	Gallon	0.6	2(ss,gl), 4(cs)
Agitator	HP	0.56	As above
Cooling tower	Gpm	0.6	1.8
Refrigeration (mech.)	Tons	0.55	1.4
Compressor	HP	0.8	3.1

Legend: ss = stainless steel, cs = carbon steel, gl = glass lined.

The approximate installed cost of the equipment may be obtained by multiplying the cost of the equipment by a factor (Lang factor) as shown in Table 6.1. The value of K and n can be estimated from the costs of two sizes. Given the cost of one size, the cost of another size may be estimated from Table 6.1.

PROBLEMS

6.1 The following table lists the variables of a heat exchanger handling fluids with no change of phase:

		Hot Side	Cold Side
Flow rate		W_1	W_2
Temperature in		T_1	T_2
Temperature out		T_3	T_4
Heat exchanged	Q		
Area	A		
Logarithmic temperature difference	LMTD		
Overall heat transfer coefficient	U		
Type of exchanger K	countercurrent		

Find the degrees of freedom for optimization if W_1, T_1, T_3, T_2, and K are fixed by the environment, and find the number of economic degrees of freedom.

6.2 You have been asked to reduce the cost of solvent losses by installing a refrigerated condenser. Your company investment guideline requires an ROI after tax greater than 25% for this type of investment. From the following data, decide whether the installation is justified or not. Depreciation of fixed investment is 10%, and the tax rate is 52%.

	Base Case	Proposed Project
Investment	0	$200,000
Annual cost		
(A) Solvent losses	$400,000/yr	$100,000/yr
(B) Power cost	$200,000/yr	$225,000/yr
(C) Cooling water	$900,000/yr	$100.000/yr
Total	$690,000/yr	$425,000/yr

6.3 The cost of a 250 m^2 exchanger is $500,000. What is the estimated order-of-magnitude cost of a similar 900 m^2 exchanger? What will be the installed cost of the 900 m^2 exchanger? Use 0.62 exponent for cost and 3.3 as the Lang factor for installation.

6.4 Formulate the equation of the total annual cost of insulating an oven of 10 m^2 heat transfer area as a function of insulation thickness, T. The following information is available:

Outside temperature:	15°C
Air film heat transfer coefficient:	23 w/m^2·K
Thermal conductivity of insulation:	0.05 w/m^2·k/mv
Installed insulation cost:	$425/m^3
Cost of heat:	$.004/10^6 J

Economic life:	10 years
Annual operating hours:	8700 hrs
Maintenance:	3% of fixed investment
Insurance and property taxes:	1.5% of fixed investment

6.5 Production of a chemical in a chemical plant of a rated manufacturing capacity 100,000 tons of chemical per year involves the following costs:

Plant investment: $10,000,000
Overhead cost: $1,000,000/year
Manufacturing cost: $50/ton
Selling price: $100/ton

Assuming a depreciation of 10%, calculate the break-even capacity as a percent of rated capacity of the chemical plant.

SOLUTIONS

6.1 The number of variables, M, given by the problem: 11.
Compute the number of independent sources of information:

$$Q = UA(\Delta T)_{LMTD}$$

$$U = \text{Function}(W, T, K)$$

$$\Delta T_{LMTD} = \frac{(T_1 - T_4) - (T_3 - T_2)}{\ln\frac{(T_1 - T_4)}{(T_3 - T_2)}}$$

$$Q = W_1 C_{P1}(T_1 - T_3)$$

$$W_1 C_{P1}(T_1 - T_3) = W_2 C_{P2}(T_4 - T_2)$$

Total number of independent sources of information = 5.
Total number of freedom for optimization = 11 − 5 = 6.
Total number of environmental degrees of freedom, by the problem, is 5.
Hence, total number of economic degrees of freedom = 6 − 5 = 1.

6.2 The justification of the project should be based on ROI of the incremental cost of the proposed project relative to the base case.

$$\text{ROI} = \frac{[(C + dI_F)_{\text{base case}} - (C + dI_F)_{\text{other case}}](1-t)}{(I_F)_{\text{other case}} - (I_F)_{\text{base case}}}$$

Hence, $\text{ROI} = \dfrac{[690{,}000 - (425{,}000 + 0.1 \times 200{,}000)](1 - 0.52)}{200{,}000}$

$$= 58.8\% > 25\%$$

Therefore, the installation is economically justified.

6.3 The estimated cost of 900 m² exchanger = $500{,}000 \times (900/250)^{0.62} = \$1{,}106{,}312$
Installed cost = $3.3 \times 1{,}106{,}312 = \$3{,}650{,}830$

6.4 Investment for insulation = (10 m²) × (T meter) × (425 $/m³) = 4,250 T($)

$$\text{Heat lost} = UA(\Delta t) = \frac{A(\Delta t)}{\frac{1}{h} + \frac{T}{k}} = \frac{(10)(293 - 15)}{0.04348 + 20T} = \frac{2780}{0.04348 + 20T} W$$

$$\text{Annual cost of heat lost} = \frac{2780 \times 8700 \times 3600 \times 10^{-9}}{0.04348 + 20T} = \frac{348.28}{0.04348 + 20T}$$

Annual depreciation = $0.1 \times 4250T = 425T$
Annual maintenance cost = $0.03 \times 4250T = 127.5T$
Annual taxes and insurance = $0.015 \times 4250T = 63.75T$

$$\text{Total annual cost} = \frac{348.28}{0.04348 + 20T} + 616.25T$$

6.5 **Given:**

Rated capacity: 100,000 tons/year
Plant investment: $10,000,000
Overhead cost: $1,000,000/year.
Manufacturing cost: $50/ton
Selling price: $100/ton
Depreciation = 10%

Assumption:

Depreciation is straight-line

Calculation:

F = Fixed cost = depreciation + overhead cost = $0.1(10,000,000) + 1,000,000$
　　= $2,000,000$/year
s = selling price = $100/ton
c = variable cost = manufacturing cost = $50/ton

$$Q_B = \frac{F}{s-c} = \frac{2,000,000(\$/yr)}{(100-50)(\$/ton)} = 40,000 \text{ tons/yr}$$

% Break-even capacity = $(40,000/100,000)100 = 40$

CHAPTER 7

Heat Transfer[1]

OUTLINE

CONDUCTION 103
Flat Single Plate ■ Composite Flat Walls ■ Cylindrical Pipe ■ Hollow Sphere

CONVECTION 105

RADIATION 105
Radiation from a Body ■ Radiation Absorbed by a Body ■ Radiation Exchange ■ Black Body ■ Gray Body

FLUID-TO-FLUID HEAT TRANSFER ACROSS A SOLID WALL 107
Log Mean Temperature Difference and Correction Factor (F_T) ■ Estimation of Outlet Temperatures ■ Heat Capacity and Specific Heat ■ Other Designs

UNSTEADY STATE HEAT TRANSFER 111
Isothermal Heating Medium ■ Isothermal Cooling Medium ■ Non-Isothermal Heating Medium ■ Non-Isothermal Cooling Medium

REFERENCES 114

The three modes of heat transfer are conduction, convection, and radiation. In this chapter we review these basic modes and their equations. We also summarize their use in several general design contexts for chemical engineering.

CONDUCTION

Conduction is a mode of heat transfer in which vibrational energy from one molecule to another migrates across a medium caused by temperature difference without intermixing or flow of material. A steady state conduction of heat is given by Fourier's law:

$$q_k = -kA(dT/dx) \qquad (7.1)$$

---
[1] Throughout this chapter, SI units are noted in brackets [].

where

q_k = heat transferred by conduction in the x direction, Btu/h[W]
k = thermal conductivity of material, Btu·ft/h·ft²·°F [W/m·K]
A = area perpendicuar to the direction of heat flow, ft²[m²]
(dT/dx) = temperature gradient in the x direction, °F/ft[K/m]

The following sections summarize the application of Fourier's law to several simple geometries.

Flat Single Plate

The rate of heat transfer through a flat single plate is given by:

$$q_k = \frac{kA(T_{hot} - T_{cold})}{X} = \frac{T_{hot} - T_{cold}}{\frac{X}{kA}} = \frac{\Delta T}{R_k} \quad (7.2)$$

where

X is the plate thickness, ft[m]
$R_k = X/kA$ is the thermal resistance, h·°F/Btu[K/W]

Composite Flat Walls

The rate of heat transfer through a series of flat plates forming a composite wall is given by:

$$q_k = \frac{T_0 - T_1}{R_{k1}} = \frac{T_1 - T_2}{R_{k2}} = \cdots = \frac{T_{n-1} - T_n}{R_{kn}} = \frac{T_0 - T_n}{R_{kt}} \quad (7.3)$$

where $T_0 > T_1 > \cdots > T_n$
R_{kn} = resistance of nth plate = $X_n/(k_n A_n)$
R_{kt} = total resistance = $(R_{k1} + R_{k2} + \cdots + R_{kn})$

Cylindrical Pipe

For cylindrical pipe, the relevant application of Fourier's law is:

$$q_k = \frac{2\pi k L(T_i - T_o)}{\ln(r_o/r_i)} = \frac{T_i - T_o}{\ln(r_o/r_i)/(2\pi k L)} = \frac{T_i - T_o}{R_k} \quad (7.4)$$

where

L = length of pipe, ft[m]
T_i = temperature at the inside wall, °F[K]
T_o = temperature at the outside wall, °F[K]
r_o, r_i = outside, inside radii of pipe, both in the same unit
$R_k = \ln(r_o/r_i)/(2kL)$, h·°F/Btu[K/W]

The heat transfer through composite cylindrical walls may be computed in the same manner as shown in equation (7.3) but using the resistance formula noted in this paragraph.

Hollow Sphere

For a hollow sphere, the rate of heat transfer is given by

$$q_k = \frac{T_i - T_o}{(r_o - r_i)/(4\pi k r_o r_i)} = \frac{T_i - T_o}{R_k} \tag{7.5}$$

where r_o, r_i are the radii of the outside and inside surfaces, both in the same unit. The heat transfer through a composite sphere may be computed in the same manner as shown in equation (7.3) but using the resistance formula noted in equation (7.5).

CONVECTION

Convection is a mode of heat transfer between a surface and a moving fluid when they are at different temperatures. When the fluid moves naturally, induced by buoyancy, the mode of heat transfer is called natural convection; otherwise it is called forced convection. It is generally assumed that the resistance to convective heat transfer lies in the boundary layer via a conductive mode. By analogy with equation (7.2)

$$q_c = kA(T_{hot} - T_{cold})/X_{boundary\ layer} = h_c A(T_{hot} - T_{cold})$$
$$= \frac{T_{hot} - T_{cold}}{1/(h_c A)} \tag{7.6}$$

where
  q_c is the convective heat transfer rate, Btu/h[W]
  h_c is the convective heat transfer coefficient, Btu/h·ft^2·°F[W/m^2·K]

The overall heat transfer coefficient, U, through composite walls, including inside and outside convection, is

$$1/U = 1/h_i + \sum X_i/k_i + 1/h_o \tag{7.6a}$$

$$q = UA\Delta T \tag{7.6b}$$

RADIATION

Radiation is a mode of heat transfer that requires no medium, in contrast with conduction and convection. Radiant energy is emitted in the form of electromagnetic waves or photons from any matter (solid, liquid, or gas) that is at a finite temperature at the expense of its internal energy.

Of the total radiant energy received by a body, a fraction (α = absorptivity) is absorbed by the body, a fraction (R = reflectivity) is reflected away from the body, and the balance (τ = transmitted) is transmitted through the body. Thus

$$\alpha + R + \tau = 1 \tag{7.7}$$

For a black body, $R = \tau = 0$, so $\alpha = 1$
For an opaque body $\tau = 0$, so $\alpha + R = 1$

Radiation from a Body

The radiant energy emitted from a surface is given by the Stefan-Boltzmann law:

$$q_{r1\to} = \sigma A_1 \varepsilon_1 T_1^4 \qquad (7.8)$$

where

$q_{r1\to}$ = heat emitted due to radiation from body 1, Btu/h[W]
σ = Stefan-Boltzmann constant = 0.173×10^{-8} Btu/(h·ft^2·°R^4)[5.67×10^{-8} W/m^{2}K^4]
A_1 = surface area of the body, ft^2[m^2]
ε_1 = emissivity of body 1 (depends on the nature of the surface), fraction
T_1 = temperature of body 1, °R[K]

Radiation Absorbed by a Body

The radiant energy absorbed by a body (1) of surface of area A_1 at temperature T_1 from black body (2) surroundings or enclosure at T_2 may be given by:

$$q_{r1-2} = \sigma A_1 \alpha_{1-2} T_2^4 \qquad (7.9)$$

where the arrow denotes the direction of the flow of radiant energy α_{1-2} is the ratio of the energy absorbed by the body (1) to the energy incident coming from the enclosure at T_2.

Radiation Exchange

Radiation exchange between a body of area A_1 and temperature T_1 and its enveloping surrounding at temperature T_2 is given as:

$$q_{r12} = \sigma A_1 \left(\varepsilon_1 T_1^4 - \alpha_{1-2} T_2^4 \right) \qquad (7.10)$$

Black Body

A black body is an ideal surface having $\alpha = \varepsilon = 1$. In a real situation, when a small object is enclosed in a large chamber, the chamber may be treated as a black body.

Gray Body

For a gray body, $\alpha = \varepsilon$ = constant and is <1 at all temperatures. Many real objects behave like gray bodies. Radiation between two close, adjacent gray surfaces, each of area A, but at temperature T_1 and T_2 may be given by:

$$q_{r12} = \frac{\sigma A \left(T_1^4 - T_2^4 \right)}{1/\varepsilon_1 + 1/\varepsilon_2 - 1} \qquad (7.11)$$

Radiation between one gray surface (A_1, T_1, ε_1) enclosed by another gray surface (A_2, T_2, ε_2) may be given by

$$q_{r12} = \frac{\sigma A_1 \left(T_1^4 - T_2^4 \right)}{1/\varepsilon_1 + (A_1/A_2)(1/\varepsilon_2 - 1)} \qquad (7.12)$$

When $A_2 \gg A_1$:

$$q_{r12} = \sigma A_1 \varepsilon_1 \left(T_1^4 - T_2^4\right) \qquad (7.13)$$

When A_1 is in ft², and T is in °R:

$$q_{r12} = 0.1713 A_1 \varepsilon_1 [(T_1/100)^4 - (T_2/100)^4], \text{Btu/h} \qquad (7.14)$$

$$h_r = \frac{0.1713 \varepsilon_1 [(T_1/100)^4 - (T_2/100)^4]}{T_1 - T_2} \qquad (7.15)$$

where, h_r is radiation heat transfer coefficient, Btu/h·ft²·°F.

When A_1 is in m², and T is in K:

$$q_{r12} = 5.67 A_1 \varepsilon_1 [(T_1/100)^4 - (T_2/100)^4], \text{W} \qquad (7.16)$$

$$h_r = \frac{5.67 \varepsilon_1 [(T_1/100)^4 - (T_2/100)^4]}{T_1 - T_2} \qquad (7.17)$$

where h_r is radiation heat transfer coefficient, W/m²·K.

FLUID-TO-FLUID HEAT TRANSFER ACROSS A SOLID WALL

The following equations present the tools necessary to calculate heat transfer from fluid to fluid separated by a cylindrical solid boundary when all the modes of heat transfer are involved with fouling resistances. Heat flows from inside to outside.

The following table explains the points of heat transfer.

Point in Heat Transfer	Representation and Remark
1	Center of pipe, hot fluid
2	Inner surface of inside fouling, h_{12} = inside convection film coefficient between points 1 & 2.
3	On inner surface of the cylindrical wall, A_3 = inside area of cylinder, f_{d23} = inside fouling resistance.
4	On outside surface of cylindrical wall. l_{34} = wall thickness, A_{mp} = mean cylinder area
5	On outside surface of outer fouling, f_{d45} = outside fouling resistance
6	Outside bulk cold fluid. h_{56} = outside convection coefficient between points 5 & 6. h_{ro} = outside radiation coefficient.

$$q = \frac{T_1 - T_2}{1/(h_{12} A_3)} = \frac{T_2 - T_3}{f_{d23}/A_3} = \frac{T_3 - T_4}{l_{34}/(k_w A_{mp})} = \frac{T_4 - T_5}{f_{d45}/A_4}$$

$$= \frac{T_5 - T_6}{1/[(h_{56} + h_r) A_4]} = \frac{T_1 - T_6}{1/(U_3 A_3)} = \frac{T_1 - T_6}{1/(U_4 A_4)} \qquad (7.18)$$

Let $T_1 - T_6 = T$, $h_{12} = h_i$, $A_3 = A_i = \pi D_i L$, $f_{d23} = f_{di}$, $1_{34} = 1_w$, $A_{mp} = \pi((D_o + D_i)/2)L = \pi D_{av} L$, $f_{d45} = f_{do}$, $A_4 = A_o = D_o L$, $h_{56} = h_o$, $U_3 = U_i$, $U_4 = U_o$:

$$Q = U_i A_i (\Delta T) = U_o A_o (\Delta T) \tag{7.19}$$

$$\frac{1}{U_i A_i} = \frac{1}{U_o A_o} \tag{7.20}$$

$$\frac{1}{U_o} = \frac{D_o}{h_i D_i} + \frac{f_{di} D_o}{D_i} + \frac{1_w D_o}{k_w D_{av}} + f_{do} + \frac{1}{h_o + h_{ro}} \tag{7.21}$$

where
- U = overall heat transfer coefficient, Btu/h·ft²·°F [W/m²·K]
- h = local convection heat transfer coefficient Btu/h·ft²·°F [W/m²·K]
- h_r = radiation heat transfer coefficient [subscript ri denotes inside, and ro outside, coefficient], Btu/h·ft²·°F [W/m²·K]
- D = diameter of cylinder ft[m]
- L = Length of pipe, ft[m]
- f_d = fouling resistance, h·ft²·°F/Btu [m²·K/W]
- 1_w = thickness of cylinder wall, ft[m]
- k_w = thermal conductivity of wall, Btu/h·ft·°F [W/m·K],
  suffix (i = inside, o = outside).

If the inside radiation is significant, then h_i may be substituted with $h_i + h_{ri}$.

When surfaces are clean, the inside coefficient is high and wall resistance is low:

$$U_o = h_o + h_{ro} \tag{7.22}$$

When surfaces are clean, wall resistance is low, the radiation coefficient is negligible, and D_o/D_i is assumed 1 for approximation:

$$U_o = (h_o)(h_i)/(h_o + h_i) \tag{7.23}$$

When radiation is neglected, the relation between the clean overall heat transfer coefficient (i.e., no fouling resistances), U_C, and service (also known as dirty) overall heat transfer coefficient, U_D, of a heat exchanger may be obtained from equation (7.21) as follows:

$$\frac{1}{U_D} = \frac{1}{U_C} + f_{di}\left(\frac{D_o}{D_i}\right) + f_{do} \tag{7.23a}$$

where

$$\frac{1}{U_c} = \frac{1}{h_i}\left(\frac{D_o}{D_i}\right) + \frac{l_w}{k_w}\left(\frac{D_o}{D_{av}}\right) + \frac{1}{h_o} \tag{7.23b}$$

Log Mean Temperature Difference and Correction Factor (F_T)

When the terminal temperature differences are not equal in an exchanger, a logarithm mean temperature difference (LMTD) is evaluated as follows.

Countercurrent flow

T_1 ----------hot fluid---------→ T_2
t_2 ←-----cold fluid---------- t_1

$\Delta T_1 = T_1 - t_2 \qquad \Delta T_2 = T_2 - t_1$

Cocurrent or parallel flow

T_1 ----------hot fluid---------→ T_2
t_1 ----------cold fluid--------→ t_2

$\Delta T_1 = T_1 - t_1 \qquad \Delta T_2 = T_2 - t_2$

$$LMTD = \frac{\Delta T_1 - \Delta T_2}{\ln\left[\frac{\Delta T_1}{\Delta T_2}\right]} \qquad (7.24)$$

Obviously, when terminal temperature differences are identical, LMTD does not apply, and the terminal temperature difference is used as the driving force.

Unless one or both of the fluids exchange heat under isothermal conditions, LMTD for countercurrent flow for the same process terminal temperatures will be greater than the LMTD for cocurrent flow, and therefore, LMTD for countercurrent flow has greater potential for heat recovery. When at least one fluid exchanges heat isothermally, LMTD for countercurrent flow is the same as that in the cocurrent flow.

In the multipass shell and tube heat exchangers, the flows are not truly countercurrent. Therefore, a correction factor, F_T, is introduced. Perry and Green provide an evaluation of this parameter. The value of F_T is 1 for a true countercurrent exchanger and is less than 1 for any other design, implying higher heat recovery by a countercurrent arrangement than any other arrangement for non-isothermal exchange of heat.

$$Q = UAF_T(\Delta T)_{lm} \qquad (7.25)$$

The correction factor F_t is determined from a graphical plot as a function of two other parameters S (or P) and R defined as follows:

$$S = \frac{t_2 - t_1}{T_1 - t_1} = \frac{\text{cold fluid temperature rise}}{\text{maximum temperature difference between hot and cold fluid}}$$

Some references label the above parameter as P. It is also known as temperature efficiency of the exchanger.

$$R = \frac{T_1 - T_2}{t_2 - t_1} = \frac{\text{hot fluid temperature drop}}{\text{cold fluid temperature rise}}$$

When R is zero, such as when there is isothermal heating medium, $F_T = 1$; when R is infinity or S is zero, such as when there is isothermal cooling medium, $F_T = 1$. Such graphs will be provided if the solution of such a problem is required.

Estimation of Outlet Temperatures

Outlet temperatures for cocurrent and countercurrent flows with no phase change, given inlet temperatures (T_1, t_1), heat capacities (C, c), overall heat transfer coefficient (U), heat transfer area (A), and flow rates (W, w), can be estimated beginning with the following equations:

$$\text{For hot fluid: } q = WC(T_1 - T_2) \tag{7.26}$$

$$\text{For cold fluid: } q = wc(t_2 - t_1) \tag{7.27}$$

$$\text{Heat balance: } q = WC(T_1 - T_2) = wc(t_2 - t_1) = UA(T)_{lm} \tag{7.28}$$

Define $R = wc/(WC)$ and $K_1 = e^{UR(R-1)/wc}$, $K_2 = e^{UA(R+1)/WC}$, then for countercurrent flow

$$T_2 = \frac{(1-R)T_1 + (1-K)Rt_1}{1-RK_1} \tag{7.29}$$

For cocurrent flow:

$$T_2 = \frac{(R+K_2)T_1 + (K_2-1)Rt_1}{(R+1)K_2} \tag{7.30}$$

For both cases, calculate t_2 from equation (7.28):

$$t_2 = (T_1 - T_2)/R + t_1 \tag{7.31}$$

$$= (T_1 - t_1)(K_1 - 1)/(RK_1 - 1) + t_1 \tag{7.31a}$$

Symbols of the preceding equations are as follows:

	Hot Fluid	Cold Fluid
Flow rate	W	w
Temperature in	T_1	t_1
Temperature out	T_2	t_2
Heat capacity	C	c

Heat Capacity and Specific Heat

These two terms are often synonymously used; their numerical values in American units and thermochemical systems are about the same. However, they are different in concept. *Specific heat* is the ratio of heat capacity of a substance to the heat capacity of an equal mass of water at a reference temperature (15 °C). Specific heat is dimensionless, but it is dependent on temperature. *Heat capacity* is the amount of heat required to raise the temperature of a body by one degree. Examples of the units of heat capacity are Btu/lb · °F and J/kg · K.

Other Designs

The techniques demonstrated through equations (7.26) to (7.31a) apply to true countercurrent and cocurrent exchangers, such as a double pipe or plate heat exchangers, or shell and tube exchangers with no crossflow. In shell and tube exchangers with baffles and multipass design, and cross-flow exchangers, those equations would give erroneous results. The problem could be solved by incorporating a correction factor into equation (7.25); however, it involves trial and error in applications where performance of a given exchanger is to be evaluated.

Another elegant solution method is the effectiveness-NTU method (Incropera and Dewitt). This method avoids trial and error solution in evaluating the performance of a given exchanger, which may be cocurrent, countercurrent, cross flow, or shell and tube type. An outline of the method involves the utilization of the following equations.

$$\text{Hot fluid heat capacity rate: } C_h = WC \quad (7.31\text{b})$$

$$\text{Cold fluid heat capacity rate: } C_c = wc \quad (7.31\text{c})$$

$$\text{Maximum possible heat transfer rate: } q_{max} = C_{min}(T_1 - t_1) \quad (7.31\text{d})$$

where C_{min} is C_h or C_c, whichever is *smaller*.

$$\text{Actual heat transfer rate: } q = \varepsilon(q_{max}) = \varepsilon C_{min}(T_1 - t_1) \quad (7.31\text{e})$$

$$\text{The number of transfer units: } NTU = (UA)/C_{min} \quad (7.31\text{f})$$

$$C_r = C_{min}/C_{max} \quad (7.31\text{g})$$

where C_{max} is C_h or C_c, whichever is *larger*.

Effectiveness, ε, may be determined from one of the following equations depending on what information is available

$$\varepsilon = \frac{C_h(T_1 - T_2)}{C_{min}(T_1 - t_1)} \quad (7.31\text{h})$$

$$\varepsilon = \frac{C_c(t_2 - t_1)}{C_{min}(T_1 - t_1)} \quad (7.31\text{i})$$

$$\varepsilon = \frac{1 - e^{-NTU[1 - C_r]}}{1 - C_r e^{-NTU[1 - C_r]}} \text{ for counter flow exchanger} \quad (7.31\text{j})$$

$$\varepsilon = \frac{1 - e^{-NTU[1 + C_r]}}{1 + C_r} \text{ for parallel flow (cocurrent) exchanger} \quad (7.31\text{k})$$

$$\varepsilon = 2\left[1 + C_r + \left(1 + C_r^2\right)^{0.5}\left(\frac{1 + e^{-NTU\left(1 + C_r^2\right)^{0.5}}}{1 - e^{-NTU\left(1 + C_r^2\right)^{0.5}}}\right)\right]^{-1} \quad (7.31\text{l})$$

for a shell and tube exchanger with one shell pass and any multiple of two tube passes. (7.31m)

For shell and tube exchangers with two shell passes and any multiple of four tube passes and cross flow exchangers, see Incropera and Dewitt.

UNSTEADY STATE HEAT TRANSFER

So far our review has focused on steady state heat transfer. We conclude this chapter with a brief look at several equations for situations of unsteady state heat transfer.

Isothermal Heating Medium

To calculate the time to heat a liquid mass in a perfectly mixed tank heated by an isothermal heating medium flowing through a coil or jacket (agitator heat or other heat effect is not considered), the equations are

$$\Theta = \frac{Mc}{UA} \ln \frac{T_1 - t_1}{T_1 - t_2} \quad (7.32)$$

$$t_\Theta = T_1 - (T_1 - t_1)e^{-k\Theta}, \quad \text{where } k = UA/(Mc) \quad (7.33)$$

$$Q_\Theta = UA(T_1 - t_\Pi) = UA(T_1 - t_1)e^{-k\Theta} \quad (7.34)$$

$$Q_{max} = UA(T_1 - t_1) \quad (7.34a)$$

Here,
- Θ = heating time, h[s]
- M = mass of liquid to be heated, lb[kg]
- c = heat capacity of the liquid to be heated, Btu/lb·°F[J/kg·K]
- U = overall heat transfer coefficient of the heating surface, Btu/h·ft²·°F [W/m²·K]
- A = heat transfer area, ft²[m²]
- T_1 = constant temperature of heating medium, °F[K]
- t = batch temperature, °F[K], subscript (1 = initial, Θ at time Θ hr[s], 2 = final, max = maximum)
- Q = instantaneous heat load, Btu/h[W]

Isothermal Cooling Medium

To calculate the time to cool a liquid mass in a perfectly mixed tank cooled by an isothermal cooling medium flowing through a coil or jacket (agitator heat or other heat effect is ignored), the equations are

$$\Theta = \frac{MC}{UA} \ln \frac{T_1 - t_1}{T_2 - t_1} \quad (7.35)$$

$$T_\Theta = t_1 + (T_1 - t_1)e^{-k\Theta}, \quad \text{where } k = UA/(MC) \quad (7.36)$$

$$-Q_\Theta = UA(T_1 - t_1)e^{-k\Theta} \quad (7.37)$$

Here,
- Θ = cooling time, h[s]
- M = mass of liquid to be cooled, lb[kg]
- C = heat capacity of the liquid to be cooled, Btu/lb·°F[J/kg·K]
- U = overall heat transfer coefficient of the heating surface, Btu/h·ft²·°F [W/m²·K]
- A = heat transfer area, ft²[m²]
- t_1 = constant temperature of cooling medium, °F[K]
- T = batch temperature, °F[K], subscript (1 = initial, Θ = at time Θ hr[s], 2 = final)
- $-Q$ = instantaneous cooling load, Btu/h[W]

Non-Isothermal Heating Medium

To calculate the time to heat a liquid mass in a perfectly mixed tank heated by a non-isothermal heating medium flowing through a coil or jacket (agitator heat or other heat effect is not considered), use the following equations:

$$\Theta = \frac{Mc}{WC} \times \frac{K}{K-1} \ln \frac{T_1 - t_1}{T_1 - t_2}$$

$$K = e^{UA/(WC)} \tag{7.38}$$

$$t_\Theta = T_1 - (T_1 - t_1)e^{-K'\Theta}, \quad \text{where } K' = \frac{WC(K-1)}{McK} \tag{7.39}$$

$$Q_\Theta = \frac{WC(K-1)(T_1 - t_1)e^{-K\Theta}}{K} \tag{7.40}$$

where
 T_1 is the inlet temperature of heating medium, °F[K]
 W is the flow rate of heating medium, lb/h[kg/s]
 C is the heat capacity of heating medium, Btu/lb·°F[J/kg·K]
 t is the batch temperature, subscript (1 = initial, Θ = at time Θ hr[s], 2 = final)
 Q = instantaneous heating load, Btu/h [W]

Non-Isothermal Cooling Medium

The relevant equations to calculate the time to cool a liquid mass in a perfectly mixed tank cooled by a non-isothermal cooling medium flowing through a coil or jacket (agitator heat or other heat effect is ignored) are

$$\Theta = \frac{MC}{wc} \times \frac{K}{K-1} \ln \frac{T_1 - t_1}{T_2 - t_1}$$

$$K = e^{UA/(wc)} \tag{7.41}$$

$$T_\Theta = t_1 + (T_1 - t_1)e^{-K'\Theta}, \quad \text{where } K' = \frac{wc(K-1)}{MCK} \tag{7.42}$$

$$-Q_\Theta = \frac{wc(K-1)(T_1 - t_1)e^{-K\Theta}}{K} \tag{7.43}$$

where
 t_1 is the inlet temperature of cooling medium, °F[K]
 w is the flow rate of cooling medium, lb/h[kg/s]
 c is the heat capacity of cooling medium, Btu/lb·°F[J/kg·K]
 T is the batch temperature, subscript (1 = initial, Θ at time Θ hr[s], 2 = final)
 $-Q$ = instantaneous cooling load, Btu/h [W]

REFERENCES

1. Das, D. K., and R. K. Prabhudesai. *Chemical Engineering License Review*, 2nd ed. Kaplan AEC Education, 2004.
2. Levenspiel, O., *Engineering Flow and Heat Transfer*, New York, Plenum Press.
3. Incropera, F. P., and D. P. Dewitt, *Fundamentals of Heat Transfer*, New York, John Wiley & Sons, 3(a) pp. 521–526.
4. Perry R. H. and D. Green, *Chemical Engineers' Handbook*, 6th ed. New York, McGraw-Hill Book Co., 1984, pp. 10–27.

PROBLEMS

7.1 A countercurrent double pipe exchanger has the following process data:

	Tube Side	Shell Side
Flow, lb/h	13,000	8,000
Temperature in °F	110	510
Heat capacity, Btu/lb·°F	1.2	0.9

Outside heat transfer area, ft^2: 400
Overall heat transfer coefficient, Btu/h·ft^2·°F: 100
Estimate the tube side outlet temperature (a) using equations (7.26) through (7.31a) and (b) by the NTU method.

7.2 A batch is to be heated in a coil-in-tank agitated vessel using saturated steam. The following data are available:

Batch size	45,000 lbs
Overall heat transfer coefficient	50 Btu/h·ft^2·°F
Heating area of coil immersed	900 ft^2
Average heat capacity of batch	0.5 Btu/lb·°F
Initial batch temperature	100°F
Steam temperature	448°F
Latent heat of steam	777 Btu/lb

Calculate the batch temperature and instantaneous steam consumption rate after 30 minutes.

7.3 The inside temperature of a composite wall is maintained at 2000°F, and the outside ambient air temperature is maintained at 70°F. The composite wall consists of three layers of materials; their thicknesses from the hotter to the colder surfaces are 12, 12, and 10 inches, respectively. The corresponding thermal conductivities are 0.4, 0.2, and 0.1 Btu-ft/h·ft^2·°F, respectively. Assume that thermal conductivities are invariant with temperature and inside heat transfer resistance is negligible. The outside air film heat transfer coefficient is 2 Btu/h·ft^2·°F. Calculate the heat loss through the composite wall and the outside surface temperature.

7.4 A heat exchanger specification sheet contains the following information:

Tube outside diameter, inch	0.75
Tube inside diameter, inch	0.62
Heat transfer surface, ft^2	91.1
Heat Exchanged, Btu/h	42,300
Corrected mean temperature difference, °F	21.9
Service overall heat transfer coefficient, Btu/h·ft^2·°F	21.2
Clean overall heat transfer coefficient, Btu/h·ft^2·°F	25.3
Shell side fouling resistance, h·ft^2·°F/Btu	0.002
Tube side fouling resistance, h·ft^2·°F/Btu	0.002

Check consistency of heat duty, and calculate the percentage of extra surface area included in the specified area.

7.5 A (1-2) countercurrent shell and tube heat exchanger is fitted with segmental baffles in the shell side. The outside surface area is 540.5 ft^2 and the overall heat transfer coefficient is 101.8 Btu/ h·ft^2·°F based on outside surface area. Evaluate the performance of the exchanger to find the outlet temperatures of the fluids. Other data are shown in the table. Assume no change of phase.

	Shell Side	Tube Side
Fluid circulated	Water	Organics
Total liquid, lb/h	198,330	70,000
Temperature in, °F	85	200
Temperature out, °F	Evaluate	Evaluate
Average heat capacity, Btu/lb·°F	1	0.5

SOLUTIONS

7.1 (a) From equation (7.29):

$$R = wc/WC = 13000 \times 1.2/8000 \times 0.9 = 2.167$$

$$K_1 = e^{UA(R-1)/wc} = e^{100 \times 400(2.167-1)/13000 \times 1.2} = 19.932$$

$$T_2 = \frac{(1-2.167)510 + (1-19.932)2.167 \times 110}{1 - 2.167 \times 19.932} = 121.06$$

From equation (7.31):

$$t_2 = (510 - 121.06)/2.167 + 110 = 289.5 \,°F$$

Alternatively, equation (7.31a) may be used, thereby avoiding calculation of T_2 and saving time:

$$t_2 = (T_1 - t_1)(K_1 - 1)/(RK_1 - 1) + t_1$$
$$= (510 - 110)(19.932 - 1)/(2.167 \times 19.932 - 1) + 110 = 289.5$$

(b) **Given**

$$W = 8000 \text{ lb/h}$$
$$C = 0.9 \text{ Btu/lb} \cdot °F$$
$$w = 13000 \text{ lb/h}$$
$$c = 1 \text{ Btu/lb} \cdot °F$$
$$T_1 = 510\,°F$$
$$t_1 = 110\,°F$$
$$U = 100 \text{ Btu/h} \cdot \text{ft}^2 \cdot °F$$
$$A = 400 \text{ ft}^2$$

Required:
Calculate t_2
Assumption:
true countercurrent flow.
Calculation:

$$C_h = WC = 8000(0.9) = 7200$$
$$C_c = wc = 13000 (1.2) = 15600$$
$$C_{min} = \text{lower of the two preceding} = 7200$$
$$C_{max} = \text{higher of the two preceding} = 15{,}600$$
$$NTU = UA/C_{min} = (100 \times 400)/7200 = 5.56$$

From equation (7.31j)

$$\varepsilon = \frac{1 - e^{-NTU[1-C_r]}}{1 - C_r e^{-NTU[1-C_r]}}$$

$$= \frac{1 - e^{-5.5556[1-0.4615]}}{1 - 0.4615 e^{-5.5556[1-0.4615]}} = 0.972$$

$$q = \varepsilon\, C_{min}(T_1 - t_1) = 0.972(7200)(510-110) = 2799360 \text{ Btu/h}$$

$$t_2 = \frac{q}{wc} + t_1 = \frac{2799360}{13000(1.2)} + 110 = 289.5\,°F$$

7.2 From equation (7.33):

$$t_\Theta = T_1 - (T_1 - t_1) e^{-k\Theta}$$

$$k = UA/Mc = 50 \times 900/45000 \times 0.5 = 2$$

$$t_\Theta = 448 - (448 - 100)e^{-2 \times 0.5} = 319.98\,°F$$

From equation (7.34):

$$Q_{\Theta=0.5} = UA(T_1 - t_1)e^{-k\Theta} = 50 \times 900(448 - 100)e^{-1} = 5760992 \text{ Btu/h}$$

Steam consumption = 5,760,992/777 = 7414.4 lb/h

7.3 From equation (7.3):

$$q_k = \frac{T_0 - T_1}{R_{k1}} = \frac{T_1 - T_2}{R_{k2}} = \frac{T_2 - T_3}{R_{k3}} = \frac{T_3 - T_a}{R_{k4}} = \frac{T_0 - T_a}{R_{k1} + R_{k2} + R_{k3} + R_{k4}}$$

where
T_0 = inside temperature of composite wall, 2000 °F
T_1 = temperature at the interface of first & second wall
T_2 = temperature at the interface of second & third wall
T_3 = temperature at the interface of third wall & air film
T_a = ambient air temperature, 70 °F

$$R_{k1} = \frac{X_1}{R_{k1}} = \frac{X_1}{k_1 A} = \frac{12/12}{0.4(1)} = 2.5 \frac{\text{h} \cdot °F}{\text{Btu}}$$

$$R_{k2} = \frac{X_2}{R_{k2}} = \frac{X_2}{k_2 A} = \frac{12/12}{0.2(1)} = 5.0 \frac{\text{h} \cdot °F}{\text{Btu}}$$

$$R_{k3} = \frac{X_3}{R_{k3}} = \frac{X_3}{k_3 A} = \frac{10/12}{0.1(1)} = 8.33 \frac{\text{h} \cdot °F}{\text{Btu}}$$

$$R_{k4} = \frac{1}{h_a A} = \frac{1}{2(1)} = 0.5 \frac{\text{h} \cdot °F}{\text{Btu}}$$

In above equations A is assumed as 1 ft².

$$q_k = \frac{T_0 - T_a}{R_{k1} + R_{k2} + R_{k3} + R_{k4}} = \frac{2000 - 70}{2.5 + 5 + 8.33 + 0.5} = 118.2 \frac{\text{Btu}}{\text{h}}$$

Outside surface temperature:

$$T_3 = q_k(R_{k4}) + T_a = 118.2(0.5) + 70 = 129.1\,°F$$

7.4 Consistency of heat load may be done using equation (7.25):
$q = UAF_T(\Delta T)_{lm}$, where $F_T(\Delta T)_{lm}$ is corrected log-mean temperature difference.
Substituting given values:
$Q = 21.2(91.1)(21.9) = 42296$ Btu/h, which checks with given heat duty within allowable tolerance of industrial practice.
Recall equation (7.23a):

$$\frac{1}{U_D} = \frac{1}{U_C} + f_{di}\left(\frac{D_o}{D_i}\right) + f_{do} \qquad (7.23a)$$

Substituting the values of the parameters of the right hand side:

$$\frac{1}{U_D} = \frac{1}{25.3} + 0.002\left(\frac{0.75}{0.62}\right) + 0.002 = 0.043945$$

$$U_D = 22.76 \; \frac{\text{Btu}}{\text{h} \cdot \text{ft}^2 \cdot °\text{F}}$$

$$\text{Required surface area} = \frac{q}{UF_T(\Delta T)_{lm}} = \frac{42300}{22.76(21.9)} = 84.9 \; ft^2$$

Actual area = 91.1 ft²

$$\text{Percent excess area} = \frac{91.1 - 84.9}{84.9}(100) = 7.3\%$$

7.5 **Given**

$$w = 198{,}330 \text{ lb/h}$$
$$c = 1 \text{ Btu/lb} \cdot °\text{F}$$
$$W = 70{,}000 \text{ lb/h}$$
$$C = 0.5 \text{ Btu/lb} \cdot °\text{F}$$
$$t_1 = 85°\text{F}$$
$$T_1 = 200°\text{F}$$
$$A = 540.5 \text{ ft}^2$$
$$U = 101.8 \text{ Btu/h} \cdot \text{ft}^2 \cdot °\text{F}$$

Find

$$t_2 \text{ and } T_2$$

Calculation

$$C_c = wc = 198{,}330 \times 1 = 198{,}330$$

$$C_h = WC = 70{,}000 \times 0.5 = 35{,}000$$

Hence, $C_{min} = 35{,}000$, $C_{max} = 198{,}330$

$$C_r = C_{min}/C_{max} = 35{,}000/198{,}330 = 0.1765$$

$$NTU = UA/C_{min} = (101.8 \times 540.5)/35{,}000 = 1.572$$

$$\varepsilon = 2\left[1 + C_r + \left(1 + C_r^2\right)^{0.5}\left(\frac{1 + e^{-NTU\left(1 + C_r^2\right)^{0.5}}}{1 - e^{-NTU\left(1 + C_r^2\right)^{0.5}}}\right)\right]^{-1}$$

$$= 2\left[1 + 0.1765 + (1 + 0.1765^2)^{0.5}\left(\frac{1 + e^{-1.575(1 + 0.1765^2)^{0.5}}}{1 - e^{-1.575(1 + 0.1765^2)^{0.5}}}\right)\right]^{-1}$$

$$= 0.739$$

$q = \varepsilon C_{min}(T_1 - t_1) = 0.739(35{,}000)(200 - 85) = 2.973 \times 10^6 \text{ Btu/h}$

$t_2 = q/C_c + t_1 = (2.973 \times 10^6)/198{,}330 + 85 = 99.988\ °\text{F}$ (cold fluid outlet)

$T_2 = T_1 - q/C_h = 200 - (2.973 \times 10^6)/35{,}000 = 115.07\ °\text{F}$ (hot fluid outlet)

CHAPTER 8

Transport Phenomena

OUTLINE

DENSITY, SPECIFIC VOLUME, SPECIFIC WEIGHT, AND SPECIFIC GRAVITY 122

VISCOSITY 122

STATIC PRESSURE, STATIC HEAD 123

STEADY, INCOMPRESSIBLE FLOW OF FLUID IN CONDUITS AND PIPES 124
Bernoulli's Theorem ■ Reynolds Number

EQUIVALENT DIAMETER FOR NONCIRCULAR CONDUITS 126

FRICTION FACTOR, NEWTONIAN FLUID 127

EQUIVALENT LENGTH OF A PIPING SYSTEM 128

RESISTANCE COEFFICIENT, K, OF A FITTING 128

DRAG COEFFICIENT 130

COMPRESSIBLE FLOW 130
Mach Number ■ Velocity of Sound ■ Critical Pressure Ratio ■ Estimation of Flow of Compressible Fluid

TEMPERATURE RISE DUE TO SKIN FRICTION UNDER ADIABATIC CONDITION 132

REFERENCES 132

Transport phenomena deal with the transfer of momentum, mass, and heat in a fluid. In this chapter, we will deal with the fundamentals of momentum transfer only.

DENSITY, SPECIFIC VOLUME, SPECIFIC WEIGHT, AND SPECIFIC GRAVITY

We begin with a brief review of several key quantities and their units (throughout this chapter, SI units are noted in brackets [ ] after equivalent non-SI units):

Density (ρ) = mass/volume, lb/ft^3 [kg/m^3]

Specific volume (v) = 1/density, ft^3/lb [m^3/kg]

Specific weight (γ) = weight/volume = mass × acceleration due to gravity/volume

$$= \rho g/g_c, \text{lbf/ft}^3 [\text{N/m}^3]$$

Specific gravity of substance(s) = density of substance/density of water at 4°C

$$= \rho(\text{lb/ft}^3)/(62.43 \text{ lb/ft}^3)$$
$$= \rho(\text{kg/m}^3)/(1000 \text{ kg/m}^3)$$

Specific gravity of gas relative to air = molecular weight of gas/28.97

The following constants are used throughout:

$$g = \text{acceleration due to gravity}$$

= 32.17 ft/s^2 [9.81 m/s^2]; this varies from place to place, and planet to planet

g_c = Newton's-law proportionality factor for the gravitational force unit

= 32.17 lb·ft/s^2·lbf [1 kg·m/s^2·N]; this is a constant factor

Specific gravity is dimensionless.
Some approximate densities at standard conditions are the following:

Air: 0.075 lb/ft^3 [1.3 kg/m^3]

Water: 62.4 lb/ft^3 [1000 kg/m^3]

Steel: 462 lb/ft^3 [7400 kg/m^3]

Mercury: 845 lb/ft^3 [13,600 kg/m^3]

VISCOSITY

Viscosity is the physical property that characterizes the flow resistance of a fluid. Analogous to thermal conductivity in heat transfer, viscosity may be regarded as momentum conductivity in viscous flow.

For Newtonian fluids, the viscosity is related by

$$\tau g_c = \mu(dV/dy) \qquad (8.1)$$

where

τ = shear stress = momentum flux = rate of momentum transfer per unit area, lbf/ft^2 [N/m^2]
g_c = 32.17 lb·ft/s^2·lbf [1 kg·m/s^2·N]
μ = viscosity, lb/ft·s [kg/m·s] = $\tau g_c/(dV/dy)$ = shear stress/shear rate
V = linear velocity, ft/s [m/s]
y = distance perpendicular to direction of flow, ft [m]
dV/dy = shear rate, 1/s

Table 8.1 Effect of Time and Shear Rate on Viscosity for Various Fluid Types

Fluid Type	Time Increase	Shear Rate Increase	n'	Example
Newtonian	No effect	No effect	1	water, air
Non-Newtonian Pseudoplastic	No effect	Decrease	$0 < n' < 1$	polymer, pigment.
Dilatant	No effect	Increase	>1	Starch slurry, clayslurry, candy
Thixotropic	Decrease	No effect		asphalt, mayonnaise
Rheopective	Increase	No effect		gypsum slurry

The viscosity of Newtonian fluids is independent of time increase and shear rate increase. The viscosity of a Newtonian fluid is sensitive to temperature. For liquids, the viscosity decreases with temperature; however, for gases, it increases according to the following:

$$\mu = \mu_0 (T/273)^n \qquad (8.2)$$

where
  μ_0 = viscosity at 273 K
  T = temperature in K
  n varies from 0.65 to 1

Except for gases at high pressure, viscosity is insensitive to pressure change.

Fluids whose viscosities do not follow Equation (8.1) are called non-Newtonian fluids. For non-Newtonian fluids, the shear stress and shear rate are related by the following equation:

$$\tau g_c = \tau_0 g_c + K'(dV/dy)n' \qquad (8.3)$$

where
  K' is the consistency index
  n' is the rheological constant.

When $\tau_0 > 0$ and $n' = 1$, then the model represents Bingham plastics. When $\tau_0 = 0$ and n' has a non-zero value other than 1, the model represents non-Newtonian fluids called power law fluids. When $\tau_0 = 0$ and $n' = 1$, the model represents Newtonian fluids.

Table 8.1 summarizes the effect on viscosity as time or shear rate increases for various fluid types.

STATIC PRESSURE, STATIC HEAD

The equation for static pressure is:

$$P_h = h\rho g/g_c = h\gamma \qquad (8.4)$$

where
  P_h = static pressure, lbf/ft^2 [N/m^2]; true on any planet, due to the column of fluid only
  h = head of fluid above the point where static pressure is to be measured = static head, ft[m] of fluid

ρ = fluid density, lb/ft³ [kg/m³]
g = 32.17 ft/s² [9.81 m/s²], variable
g_c = 32.17 lb·ft/s²·lbf [1 kg·m/s²·N]
γ = specific weight, lbf/ft³ [N/m³]

Special cases are

$$P_h = 0.433\, hs \tag{8.4a}$$

where
P_h = static pressure, psi, generally on the surface of the earth, due to the column of fluid only
h = static head, ft of fluid
s = specific gravity, dimensionless

$$P_h = 9.81\, hs \tag{8.5}$$

where
P_h = static pressure, kN/m², generally on the surface of the earth
h = static head, m of fluid
s = specific gravity, dimensionless

Generally, the pressure measured by a gage is relative to the barometric pressure, and is denoted by psig, whereas the absolute pressure is denoted by psia. Thus:

$$\text{Absolute pressure} = \text{Barometric pressure} \pm \text{gage pressure} \tag{8.5a}$$

The minus sign is used for the vacuum gage.

STEADY, INCOMPRESSIBLE FLOW OF FLUID IN CONDUITS AND PIPES

The shear stress at the wall of a pipe, τ_w, due to flow of any fluid, Newtonian or non-Newtonian, is given by:

$$\tau_w = 0.25 D(\Delta P/L) \tag{8.6}$$

where ΔP = pressure drop due to skin friction, lbf/ft² [N/m²], in a pipe of length L ft [m] and internal diameter D ft [m]

Now, $\Delta P/L$ for laminar flow of Newtonian fluid in a pipe is related to viscosity, velocity or flow rate, and internal diameter as follows:

$$\Delta P/L = 32 V \mu/(g_c D^2) \text{ [Hagen-Poiseuille Equation]} \tag{8.7}$$

It follows from the two equations mentioned previously that the wall shear stress for laminar flow of Newtonian fluid may be represented by

$$\tau_w = (\mu/g_c)(8V/D) = (\mu/g_c)[32Q/(\pi D^3)] \tag{8.8}$$

where Q is the flow rate ft³/s [m³/s]. In general, for laminar or turbulent flow of Newtonian fluids:

$$\tau_w = f\rho V^2/(2 g_c) \tag{8.9}$$

For a power-law fluid

$$\tau_w = 0.25D(\Delta P/L) = K'(8V/D)^{n'} \quad \text{(8.9a)}$$

For a power-law fluid, a plot of $0.25D(\Delta P/L)$, lbf/ft^2[Pa] vs. $8V/D$, s^{-1}, on a log-log graph would provide the rheological constant, n', the slope, and consistency index, K', lbf/ft^2[Pa], at the y-axis corresponding to $8V/D = 1$ at the x-axis.

The generalized equation for frictional pressure drop ΔP, lbf/ft^2[Pa] of a power-law fluid for laminar flow may be given by:

$$\Delta P/L = \frac{32K8^{n'-1}V^{n'}}{D^{n'+1}} \quad \text{(8.9b)}$$

where velocity, V, is in ft/s[m/s], and diameter, D, is in ft[m]. For handling Bingham plastics and power-law fluids in turbulent flow, see Sultan

The velocity distribution for laminar flow in a circular tube or between planes is given by

$$V_{\text{local},r} = V_{\text{max}}[1 - (r/R)^2] \quad \text{(8.10)}$$

where

$V_{\text{local},r}$ = local velocity at a distance r, ft/s[m/s]
V_{max} = velocity at the center of the duct or planes, ft/s[m/s]
r = distance from the center line, ft[m]
R = radius of the tube or half the distance between the parallel planes, ft[m]
V_{max} = $1.22V$ for turbulent flow ($N_{Re} > 10{,}000$)
  = $2V$ for circular tubes in laminar flow
  = $1.5V$ for parallel planes, laminar flow, where V = average velocity.

Bernoulli's Theorem

Bernoulli's Theorem can be expressed as

$$Z_A(g/g_c) + V_A^2/(2\alpha g_c) + P_A/\rho_A + W = Z_B(g/g_c) + V_B^2/(2\alpha g_c) + P_B/\rho_B + F \quad \text{(8.11)}$$

where the subscript A denotes upstream conditions subscript B denotes downstream conditions, that is, the flow is from A to B

$\alpha = 1/2$ for laminar flow, 1 for turbulent flow
Z = static head, ft[m]
V = velocity, ft/s[m/s]
P = pressure, lb/ft^2[N/m^2]
W = work done on the system, ft·lbf/lb[J/kg]
F = pressure drop due to skin friction between points A and B, ft·lbf/lb[J/kg]
ρ = density, lb/ft^3[kg/m^3], on earth $g/g_c \sim 1$ lbf/lb [9.81 N/kg]

Each term of the equation represents energy per unit mass of the fluid, ft·lbf/lb [J/kg].

Reynolds Number

The dimensionless Reynolds number, N_{Re}, is defined as

$$N_{re} = DV\rho/\mu = DV/\nu = DG/\mu$$

Table 8.2 Dimensional Equations for Reynolds Number and Pressure Drop

Reynolds Number	Frictional Pressure Loss (Incompressible Newtonian Fluid)	
$N_{Re} = 123.9\, dV\rho/\mu$	$\Delta P = 0.005176\, fL\rho V^2/d$	(8.12)
$N_{Re} = 50.65\, Q\rho/(\mu d)$	$\Delta P = 0.000864\, fL\rho Q^2/d^5$	(8.13)
$N_{Re} = 379\, \rho q/(\mu d)$	$\Delta P = 0.0484\, fL\rho q^2/d^5$	(8.14)
$N_{Re} = 6.32\, W/(\mu d)$	$\Delta P = 0.0000134\, fLW^2/(\rho d^5)$	(8.15)

where
  D = diameter of pipe, ft [m]
  V = velocity, ft/s [m/s]
  ρ = density of fluid, lb/ft³ [kg/m³]
  μ = viscosity of fluid, lb/ft·s [kg/m·s]
  ν = kinematic viscosity = μ/ρ, ft²s [m²/s]
  G = mass velocity, lb/ft²·s [kg/m²·s]

Use the following conversion factor, when necessary, to convert viscosity in centipoises (cP) to other units:

$$1\, cP = 6.72 \times 10^{-4}\, lb/ft \cdot s = 0.001\, kg/m \cdot s$$

$N_{Re} < 2100$ indicates viscous (laminar) flow, $N_{Re} > 4000$ indicates turbulent flow, $2100 < N_{Re} < 4000$ indicates transition region. The upper boundary of laminar flow is given by $(N_{Re})_{critical} = 2100$. Table 8.2 summarizes several key equations for Reynolds number and pressure drop.

For the equations in Table 8.2
  d = internal diameter, in.
  f = Fanning friction factor, dimensionless
  L = pipe length, ft
  ΔP = frictional pressure loss, psi
  q = flow rate, ft³/min
  Q = flow rate, gpm
  N_{Re} = Reynolds number, dimensionless
  V = velocity, ft/s
  W = flow rate, lb/h
  μ = viscosity, centipoise, cP
  ρ = density, lb/ft³

EQUIVALENT DIAMETER FOR NONCIRCULAR CONDUITS

The equivalent diameter, D_e, is related to the hydraulic radius, r_h, by

$$D_e = 4r_h$$

where

$$r_h = \frac{\text{cross-sectional area of flow}}{\text{wetted perimeter of channel}}$$

For the annular space of concentric pipes:

$$D_e = D_2 - D_1$$

where
D_2 = inside diameter of outer pipe
D_1 = outside diameter of inner pipe

FRICTION FACTOR, NEWTONIAN FLUID

For laminar flow of a Newtonian fluid ($N_{Re} < 2100$), the friction factor is $f = 16/N_{Re}$. For turbulent flow, Figure 8.1 illustrates the relationship between the Fanning friction factor and the Reynolds number.

For fully turbulent flow, f_T is independent of Reynolds number:

$$f_T = \frac{1}{16\left[\log\left(\frac{3.7(D)}{\varepsilon}\right)\right]^2} \quad (8.16)$$

A good rule of thumb: When the Reynolds number in a pipe is equal to or greater than nominal pipe size (10^6), the flow is likely to be fully turbulent.

For non-fully turbulent flow, f is dependent on Reynolds number:

$$f = \frac{1}{2.444[\ln(0.135\,\varepsilon/D) + 6.5/N_{Re}]^2} \quad (8.16a)$$

where
D = inside diameter of pipe
ε = roughness factor of pipe, both in the same unit of length

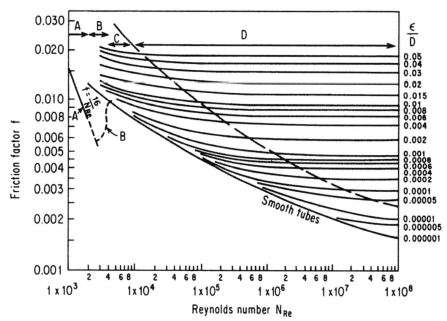

Figure 8.1 Fanning friction factor versus Reynolds number. (A) Laminar region, (B) critical region, (C) transition region, and (D) region of complete turbulence.

For smooth pipe, such as drawn tubing, glass pipe, and so on, $N_{Re} < 10^5$:

$$f = 0.079 \, N_{Re}^{-1/4} \tag{8.17}$$

This equation may also be used for turbulent flow of non-Newtonian fluids.

Caution! Before using any equation or friction factor chart for pressure drop calculation, make sure that the friction factor is consistent with the pressure drop equation. The Fanning friction factor used in the preceding is one-fourth of the Moody friction factor.

EQUIVALENT LENGTH OF A PIPING SYSTEM

The total equivalent length of a piping system, L_e, is the sum of straight piping and the equivalent lengths of fittings such as elbows, valves, and other hardware.

$$L_e = L_{\text{straight pipe}} + \Sigma (L_e)_{\text{fittings}} \tag{8.18}$$

RESISTANCE COEFFICIENT, K, OF A FITTING

The resistance coefficient, K, of a fitting is the number of velocity heads lost because of the fitting. See Table 8.3 for the L_e/D ratio and Table 8.4 for the resistance coefficient K for various fittings. Multiply L_e/D by the corresponding diameter to get the equivalent length of the fitting.

For flows that are not fully turbulent or laminar, the L/D ratio should be corrected for the friction factor. L/D varies inversely as the friction factor. The K factors of the fittings in Table 8.3 are also available as a function of fully turbulent friction factor of a given size. To find the K factor, one needs to know the fully turbulent friction factor for the size. In general, the K factor of a *fitting or a valve* of a given size is constant for all flow conditions, laminar or turbulent.

Table 8.3 Equivalent Lengths (L/D) of Valves and Fittings for Full Turbulent Flows

Valve or Fitting	Comment	L/D
Gate valve (disk or plug)	Fully open	13
Globe valve	Wing or pin guided disk	450
Angle valve	Wing or pin guided disk	200
Ball valve	Fully open & full port	13
Butterfly valve	Size 2–8 inch	45
	Size 10–14 inch	35
	Size 16–24 inch	25
Plug valve	Straight-way	18
	3-way (straight)	30
	3-way (branch)	90
Swing check valve	Tilting seat	100
Lift or stop check	Globe lift or stop	450
	Angle lift or stop	200
Standard flanged elbow	90 degree	30
	45 degree	16
Standard flanged tee	Flow through run	20
	Flow through branch	60

Table 8.4 K Factors for Miscellaneous Obstructions

Type	Comment	K
Pipe entrance	Inward projection	0.78
Pipe entrance	Sharp-edged	0.50
Pipe entrance	Slightly rounded	0.23
Pipe entrance	Well rounded	0.04
Exit from pipe	Projecting, sharp edged, or rounded	
	(a) Fluid exits to a confined space	(a) 1.0
	(b) Fluid exits to an unconfined space (see note below)	(b) 0
Sudden contraction, based on small diameter	β = small diameter/large diameter	$0.5(1 - \beta^2)$
Sudden contraction, based on small diameter	β = small diameter/large diameter	$(1 - \beta^2)^2$

Note: When the fluid exits to a confined space, the velocity goes to zero, and consequently, the kinetic energy dissipates as friction in the mixing process. In this case, $K = 1.0$. When the fluid exits to an unconfined space, such as atmosphere, the velocity of the exiting fluid in the form of a jet is maintained the same as that of the fluid inside the pipe. In this case, $K = 0$. Many engineers make a conservative error in the latter case by assuming $K = 1.0$.

When the L_e/D ratio for a specific diameter is not available, it can be estimated from the known L_e/D ratio corresponding to a diameter for fully turbulent flows as follows:

$$(L_e/D)_a = (L_e/D)_b (D_a/D_b)^4 \tag{8.19}$$

In Equation 8.19, subscript a refers to the ratio required for a specific diameter D_a, and subscript b refers to known data. It follows that the equivalent lengths of pipes of two different diameters for fully turbulent flows are related by the following:

$$(L_e)_a = (L_e)_b (D_a/D_b)^5 \tag{8.20}$$

When the Reynolds number falls below the fully turbulent region, the L_e/D ratio should be corrected for the friction factor, f, as follows:

$$\left(\frac{L}{D}\right)_1 = \left(\frac{L}{D}\right)_t \left(\frac{f_t}{f_1}\right) \tag{8.21}$$

where subscript 1 refers to an L_e/D ratio for a Reynolds number below the fully turbulent region subscript t refers to an L_e/D ratio for fully turbulent conditions as shown in Table 8.3.

Note that L/D factors for laminar flows will be lower than those for turbulent flows.

K and L_e/D are related by the following:

$$K = 4fL_e/D \tag{8.22}$$

The K factors for fittings and valves are independent of friction factors and are constant for a given size of a fitting or a valve for all flow conditions, turbulent or laminar.

The K factors should be corrected for internal diameter when the factor for a given size is known, as follows:

$$K_a = K_b (D_d/D_b)^4 \tag{8.23}$$

where K_a refers to the K factor corresponding to internal diameter D_a and K_b is the K factor for internal diameter D_b.

The frictional pressure drop, K, and L_e/D are related by

$$\frac{\Delta P}{\rho} = \left[\Sigma K + 4f\Sigma\left(\frac{L_e}{D}\right)\right]\left(\frac{V^2}{2g_c}\right) \qquad (8.24)$$

where ΔP is the frictional pressure drop in lbf/ft² [N/m²]
ρ is the density in lb/ft³ [kg/m³]
L_e and D are the length and diameter, respectively, of the fitting in the same unit
V is the velocity in ft/s [m/s]
$g_c = 32.17$ lb·ft/s²·lbf [1 kg·m/s²·N]

DRAG COEFFICIENT

When a fluid flows past an immersed body, the force in the direction of flow exerted by the fluid on the immersed body is called drag, F_D, lbf [N]. If A_p is the projected area, ft² [m²], of the immersed body on a plane perpendicular to the direction of flow, then the drag coefficient, C_D (dimensionless), analogous to the friction factor, f, is given by the following:

$$F_D/A_p = C_D \rho V^2/2g_c \qquad (8.25)$$

where V is the relative velocity between the immersed body and the fluid.

By Newton's third law of motion, an equal and opposite force is exerted by the immersed body on the fluid. Graphical plots of C_D vs. the Reynolds number ($D_p V \rho/\mu$) are available in Perry (pp. 5–64), where D_p is the diameter, ft [m], of the spherical particle having the same volume as the particle.

COMPRESSIBLE FLOW

A fluid is compressible when the density changes with pressure. For practical applications, when the high density/low density ratio is roughly >2, compressible flow should be considered in evaluating flow rates and pressure drop.

Mach Number

$$N_{Ma} = V/c$$

where V = velocity of fluid, and c = velocity of sound in the same fluid under identical pressure and temperature, both velocities being in the same unit $N_{Ma} > 1$ indicates supersonic flow. $N_{Ma} = 1$ indicates sonic flow. $N_{Ma} < 1$ indicates subsonic flow.

In practical application of compressible fluid flow, when the downstream pressure is lowered with a fixed upstream pressure and a fixed uniform geometry, the velocity of the compressible fluid increases and so does the flow until the velocity reaches the velocity of sound in the fluid ($N_{Ma} = 1$). If the downstream pressure is lowered further, flow does not increase. This is also known as chocked flow. Thus the knowledge of Mach number allows one to determine the maximum possible flow for a given inlet pressure and geometry of the hardware for a compressible fluid. All compressible flows, whether subsonic or supersonic initially in a pipe of uniform diameter, will reach sonic velocity as the length is increased. However, a subsonic flow can be transformed into a supersonic flow through a convergent-divergent nozzle if the sonic velocity is reached at the throat and the pressure at the outlet is lower than the throat pressure.

Velocity of Sound

The equation for the velocity of sound is given as

$$c = (kg_c RT/M)^{1/2} \text{ for an ideal gas, } k = C_p/C_v \quad (8.25a)$$

where
- c = velocity of sound in an ideal gas, ft/s [m/s]
- C_p = heat capacity at constant pressure
- C_v = heat capacity at constant volume
- g_c = 32.2 lb·ft/s^2·lbf [1 kg·m/s^2·N]
- R = 1545 ft·lbf/lb·mol °R [8314 N·m/kmol·K]
- M = molecular weight of gas, lb/lbmol [g/mol]

Critical Pressure Ratio

For a constant inlet pressure, the flow rate of a compressible fluid increases as the downstream pressure is lowered until a downstream pressure is reached when the flow rate reaches maximum and the fluid velocity reaches sonic velocity. At this condition, the ratio of the downstream pressure to upstream pressure in absolute scale is called the *critical pressure ratio*. For an ideal gas

$$\frac{P_{\text{downstream}}}{P_{\text{upstream}}} = \left[\frac{2}{k+1}\right]^{\frac{k}{k-1}} \quad (8.26)$$

If downstream pressure is less than what is calculated from the preceding equation, the flow rate becomes independent of the downstream pressure. As a rule of thumb, when the preceding ratio is less than 0.5, critical flow (choked flow) is anticipated.

Estimation of Flow of Compressible Fluid

The method for estimating flow of compressible fluid varies depending on the pressure drop.

1. When the pressure drop is <10% of inlet pressure, or at a very high pressure, use Bernoulli's equation using either inlet or outlet density.

2. When the pressure drop is >10% but <40% of inlet pressure, use Bernoulli's equation, but use the average density based on inlet and outlet conditions.

3. When the pressure drop is >40% of inlet pressure, or for any compressible flow, the following method may be used under adiabatic conditions for known upstream pressure, P_1, lbf/ft^2 [N/m^2]; upstream temperature, T_1, °R [K]; equivalent pipe length, L, ft [m]; inside diameter of pipe, D, ft [m]; heat capacity ratio, k; gas constant, R, 1545, ft·lbf/lbmol·°R [8314 N·m/kmol·K]; and molecular weight, M, lb/lbmol [g/mol].

Calculate the friction factor using Equation (8.16), and then estimate the upstream Mach number ($Ma1$) by trial and error using Equation (8.27), followed by the calculation of the remaining parameters.

$$\frac{(k+1)}{2}\ln\frac{2+(k-1)N_{Ma1}^2}{(k+1)N_{Ma1}^2} - \left(\frac{1}{N_{Ma1}^2} - 1\right) + k(4fL/D) = 0 \quad (8.27)$$

$$Y_1 = 1 + \frac{(k-1)N_{Ma1}^2}{2} \quad \text{dimensionless} \quad (8.28)$$

$$T_{choked} = 2Y_1T_1/(k+1), \,°R\,[K] \quad (8.29)$$

$$P_{choked} = P_1 N_{Ma1}[2Y_1/(k+1)]^{0.5}, \,lbf/ft^2\,[N/m^2] \quad (8.30)$$

$$G_{choked} = P_{choked}[kg_cM/(RT_{choked})]^{0.5}, \,lb/ft^2 \cdot s\,[kg/m^2 \cdot s] \quad (8.31)$$

$$\text{Mass flow} = G_{choked}\pi D^2/4, \,lb/s\,[kg/s] \quad (8.32)$$

TEMPERATURE RISE DUE TO SKIN FRICTION UNDER ADIABATIC CONDITION

Assuming skin friction is completely converted to heat, the temperature rise, $\Delta T, °F[K]$, may be given by

$$\Delta T = \Delta P/(\rho \eta C_p) \quad (8.33)$$

where
ΔP is skin frictional pressure drop, $lbf/ft^2\,[N/m^2]$
ρ is the density, $lb/ft^3\,[kg/m^3]$
$\eta = 778$ ft·lbf/Btu [1 N·m/J]
C_p is the heat capacity, Btu/lb·°F[J/kg·K]

For the fundamentals of manometers, pump hydraulics, flow meters, and compression equipment, see Das and Prabhudesai.

REFERENCES

1. Das, D. K., and R. K. Prabhudesai, *Chemical Engineering License Review*, 2nd ed., Kaplan AEC Education, 2004.
2. Levenspiel, O., *Engineering Flow and Heat Transfer*, New York, Plenum Press,
3. Sultan, A. A., Chemical Engineering, Dec. 19, 1988, pp. 140–146.

PROBLEMS

8.1 Find the velocity of sound in a gas of density 1 kg/m^3 at a pressure of 1.013E5 N/m^2 with the heat capacity ratio of 1.4, assuming the gas is ideal.

8.2 The rheological constant of some power-law fluid, n' is 0.586.

The existing pipe with an internal diameter of 0.154 m is to be replaced with a new pipe of the same length to reduce the pressure drop by 50% and maintain the same volumetric flow rate. Assuming the flow is laminar, what size pipe is needed?

8.3 Water flows without change of phase through a composite pipe system comprising two steel pipe sections in series: the first section has 6.065 inch inside diameter with 1000 ft of equivalent length (including the expansion effect), and the second section has 10.02 inch inside diameter with 1000 ft of equivalent length. A differential pressure of 15 psi is available between the upstream and downstream ends of the composite piping system. Assume density of water as 62.4 lb/ft^3 and viscosity as 1 cP. The roughness factor of steel pipe is 0.00015 ft. Estimate the flow through the composite section in gallons per minute using a single iteration.

For preliminary estimate use the following for the Fanning friction factor for fully turbulent flow:

$$f_T = \frac{1}{16\left[\log\left(\frac{3.7(D)}{\varepsilon}\right)\right]^2}$$

For iteration use the following for the Fanning friction factor:

$$f = \frac{1}{2.444[\ln(0.135\,\varepsilon/D) + 6.5/N_{Re}]^2}$$

Where and D are pipe roughness factor and internal pipe diameter in the same unit.

8.4 In Problem 8.3, replace the 10.02-inch pipe with an equivalent length of 6.065-inch pipe, and estimate the flow through the composite pipe.

SOLUTIONS

8.1 From Equation (8.25a)

$$c = (kg_c RT/M)^{0.5} = (kg_c P/\rho)^{0.5}$$

$$= \left(1.4 \times \frac{1 kg \cdot m}{s^2 N} \times \frac{1.013 \times 10^5 N}{m^2} \times \frac{m^3}{1 kg}\right)^{0.5} = 376.6 \text{ m/s}$$

8.2 From Equation (8.9b), for the same power-law fluid through the same pipe length:

$$\Delta P \propto V^{n'}/D^{n'+1} \propto Q^{n'}/D^{3n'+1}$$

Therefore, for the same volumetric flow rate Q through the same pipe length

$$\Delta P_1/\Delta P_2 = 2 = (D_2/D_1)^{2.758}.$$

Hence, $D_2 = D_1 \left[10^{(\log 2)2.758}\right] = 0.154 = [10^{0.1091}] = 0.198$ m

8.3 **Given**

First pipe section:
Internal diameter, $d_1 = 6.065$ inch
Equivalent length, $L_1 = 1000$ ft
Second pipe section:
Internal diameter, $d_2 = 10.02$ inch
Equivalent length, $L_2 = 1000$ ft
Pipe roughness factor = 0.00015 ft
Water density = 62.4 lb/ft^3
Water viscosity = 1 cP
Terminal pressure drop, $\Delta P = 15$ psi

Assumption
Neglect the difference of velocity heads in the two sections.

To find
Flow in gpm.

Solution
Calculate preliminary friction factor:
For the first pipe section:

$$f_1 = \frac{1}{16\left[\log\left(3.7 \frac{d}{12(\varepsilon)}\right)\right]^2}$$

$$= \frac{1}{16\left[\log\left(\frac{3.7 \times 6.065}{12 \times 0.00015}\right)\right]^2} = 0.003726$$

For the second pipe section:

$$f_2 = \frac{1}{16\left[\log\left(3.7\dfrac{d}{12(\varepsilon)}\right)\right]^2}$$

$$= \frac{1}{16\left[\log\left(\dfrac{3.7 \times 10.02}{12 \times 0.00015}\right)\right]^2} = 0.003359$$

Preliminary flow in gpm:

$$Q_P = \left[\frac{\Delta P}{0.000864(\rho)\left[\dfrac{f_1(L_1)}{d_1^5} + \dfrac{f_2(L_2)}{d_2^5}\right]}\right]^{0.5}$$

$$= \left[\frac{15}{0.000864(62.4)\left[\dfrac{0.003726(1000)}{6.065^5} + \dfrac{0.003359(1000)}{10.02^5}\right]}\right]^{0.5}$$

$$= 755.647 \text{ gpm}$$

Now iterate the friction factor:

$$\mu = 1, \quad Re_1 = \frac{50.65(Q_p)\rho}{\mu(d_1)} = \frac{50.65(755.647)62.4}{1(6.065)} = 3.938 \times 10^5$$

$$f_{1\text{rev}} = \frac{1}{2.444\left[\ln\dfrac{0.135(12)\varepsilon}{d_1} + \dfrac{6.5}{Re_1}\right]^2}$$

$$= \frac{1}{2.444\left[\ln\dfrac{0.135(12)(0.00015)}{6.065} + \dfrac{6.5}{3.938\,(10^5)}\right]^2} = 3.991 \times 10^{-3}$$

$$\mu = 1, \quad Re_2 = \frac{50.65(Q_p)\rho}{\mu(d_2)} = \frac{50.65(755.647)62.4}{1(10.02)} = 2.384 \times 10^5$$

$$f_{2\text{rev}} = \frac{1}{2.444\left[\ln\dfrac{0.135(12)\varepsilon}{d_2} + \dfrac{6.5}{Re_2}\right]^2}$$

$$= \frac{1}{2.444\left[\ln\dfrac{0.135(12)(0.00015)}{10.02} + \dfrac{6.5}{2.384\,(10^5)}\right]^2} = 3.623 \times 10^{-3}$$

Flow after the first iteration of friction factors:

$$Q_{final} = \left[\frac{\Delta P}{0.000864(\rho)\left[\frac{f_{1rev}(L_1)}{d_1^5} + \frac{f_{2rev}(L_2)}{d_2^5}\right]} \right]^{0.5}$$

$$= \left[\frac{15}{0.000864(62.4)\left[\frac{0.003991(1000)}{6.065^5} + \frac{0.003623(1000)}{10.02^5}\right]} \right]^{0.5}$$

$$= 729.904 \text{ gpm}$$

8.4 Assumption for calculating equivalent length:
The friction factor in the equivalent pipe is the same as the base pipe.

Calculation:
Calculate the equivalent length of 10.02 inch diameter pipe in terms of the 6.065 inch diameter pipe:

$$(L_e)_{6.065} = (L_e)_{10.02}\left[\frac{6.065}{10.02}\right]^5 = 1000\left[\frac{6.065}{10.02}\right]^5 = 81.25 \text{ ft}$$

Total equivalent length in 6.065 inch diameter pipe = 1000 + 81.25 = 1081.25 ft
Preliminary estimate of friction factor [see solution of Problem 8.3]:

$$f = 0.003726$$

$$Q_P = \left[\frac{\Delta P}{0.000864\ (\rho)\left[\frac{f_1(L_1)}{d_1^5}\right]} \right]^{0.5}$$

$$= \left[\frac{15}{0.000864(62.4)\left[\frac{0.003726(1081.25)}{6.065^5}\right]} \right]^{0.5}$$

$$= 752.82 \text{ gpm}$$

Using this flow, refine the friction factor:

$$\mu = 1, \text{ Re}_1 = \frac{50.65(Q_p)\rho}{\mu(d_1)} = \frac{50.65(752.82)62.4}{1(6.065)} = 3.923 \times 10^5$$

$$f_{rev} = \frac{1}{2.444\left[\ln\frac{0.135(12)\varepsilon}{d_1} + \frac{6.5}{\text{Re}_1}\right]^2}$$

$$= \frac{1}{2.444\left[\ln\frac{0.135(12)(0.00015)}{6.065} + \frac{6.5}{3.923(10^5)}\right]^2} = 3.991 \times 10^{-3}$$

Revise flow rate using the preceding friction factor:

$$Q_P = \left[\frac{\Delta P}{0.000864(\rho)\left[\frac{f_{rev}(L_1)}{d_1^5}\right]} \right]^{0.5}$$

$$= \left[\frac{15}{0.000864(62.4)\left[\frac{0.003991(1081.25)}{6.065^5}\right]} \right]^{0.5}$$

$$= 727.39 \text{ gpm}$$

CHAPTER 9

Process Control

OUTLINE

CONTROL SYSTEMS 139
Open Control Systems ■ A Closed Loop or Feedback System ■ Feed-Forward Control ■ Cascade Control ■ Servo-Operation ■ Regulatory Operation ■ On-Off (Bang-Bang) Controller ■ Proportional Control ■ Proportional-Plus-Integral Controller ■ Proportional-Plus-Derivative Control

BLOCK DIAGRAMS 142

TRANSFER FUNCTIONS 142

PROPORTIONAL BAND 143

TRANSIENT RESPONSE OF CONTROL SYSTEMS 144

FIRST AND SECOND ORDER SYSTEMS 145

REFERENCES 146

A chemical process plant is a combination of unit operations equipment, such as reactors, distillation columns, absorbers, and evaporators, and auxiliary equipment, such as tanks and heat exchangers. The basic objective of a plant is to produce marketable profit-making products from certain raw materials in an economical way. In doing so, the plant also has to meet additional requirements, such as product quality, safety of operation, and environmental regulations, and keep within the limitations of the process equipment. These requirements make it necessary to continuously monitor or control plant operations. This chapter briefly reviews the basics of process control systems.

CONTROL SYSTEMS

A control system is characterized by an output variable (for example, temperature) that is automatically controlled through the manipulation of the inputs. Automatic control is required (1) for precise control of the process to produce more uniform and high quality product, (2) for processes that are too rapid for manual control, (3) in hazardous operations where remote control is a necessity, (4) to maintain stability, and (5) for optimization of the process. Following are brief descriptions of some types and components of control systems.

Open Control Systems

The inputs to the process in an open control system are regulated independently without using the measurement of the controlled output variable to readjust the inputs.

A Closed Loop or Feedback System

A closed loop or feedback system implies that the measurement of the controlled variable is used to manipulate one of the process variables.

Feed-Forward Control

In a feed-forward control system, the measurement of one input variable is used to manipulate another input variable.

Cascade Control

In a cascade control system, the output of a primary controller is used to adjust the set point of a secondary controller. Cascade control is commonly used in feedback control. The *ratio and selector controls* are two other control modes that use two or more interconnected instruments.

Servo-Operation

In servo-operation mode, the control system is designed to follow the changes in the set point as closely as possible according to some prescribed function.

Regulatory Operation

In regulator operation, the control system is designed to keep output constant, that is, to maintain the controlled variable at a fixed value in spite of the changes in load. This is more common in the control of chemical processes.

On-Off (Bang-Bang) Controller

In this type of system, when the measured variable is below the set point, the controller is on with maximum output. However, the controller is off and the output is zero if the measured variable is above the set point. Modifications of this controller include dead band, also called differential gap, and hysteresis. On-off control action is inherently cyclic in nature.

Proportional Control

Proportional control is also called throttling control. In this control mode, the measurement of the controlled variable is matched against a set point and the error is used to determine the position of the final operator. One drawback of this control is that it can reduce the effect of load change but cannot eliminate it completely, which gives rise to a steady state difference called offset between the measurement and the set point. The basic equation of the proportional control is as follows:

$$P = \frac{100}{p} e + b$$

where
 P = proportional band, %
 e = fractional change in error or deviation of measurement from the set point
 b = output bias
 p = fractional change in controller output or manipulated variable

Gain of the controller is the change in its output dp caused by deviation de in the measurement and is expressed as

$$\frac{dp}{de} = K_C G_C$$

The gain of the controller has a steady state component K_c, which has a value of $100/P$ and a dynamic component G_c. The condition for undamped oscillation to persist is

Gain product of all elements in the loop = $K_c G_c K_p G_p = 1.0$

If the gain product < 1, oscillation will be dampened. Usually a value of 0.5 or less is desirable to avoid exceeding unity, which is the limit of stability. Quarter amplitude damping will be achieved by reducing the loop gain to 0.5.

Proportional-Plus-Integral Controller

To remove the offset of a proportional control, an integral action is added to the proportional controller. This integral action integrates the difference between the measurement and the set point and causes the controller's output to change until the error is reduced to zero. Integral action however results in an oscillatory response. Proportional-plus-integral controller is represented by the relation

$$p = \frac{100}{P}\left(e + \frac{1}{\tau_i}\int e\, dt\right)$$

where
 τ_i = integral time

Proportional-Plus-Derivative Control

Derivative action is used to increase the rate of correction. This controller is mathematically represented by the equation

$$p = \frac{100}{P}\left(e + \tau_D \frac{de}{dt}\right)$$

where
 τ_D = derivative time constant

Figure 9.1 Components of a control system block diagram:
(a) transfer function (dynamic relationship);
(b) comparison of signals

BLOCK DIAGRAMS

Block diagrams are used to show functional relationship between the various parts of the control system. The typical components of a block diagram are (a) transfer functions and (b) comparators. These are shown in Figure 9.1. They are obtained directly from the physical system, which is divided into sections whose inputs and outputs are distinctly identifiable.

TRANSFER FUNCTIONS

Control system analysis involves the determination of the transfer functions by the application of either mass, energy, or force balance to each part of the system. Developed linear differential equations are solved using Laplace transforms (Das and Prabhudesai, Wylie, and Perry and Green). A transfer function is defined as follows:

$$\text{Transfer function} = \frac{\text{Laplace Transform of Output}}{\text{Laplace Transform of Input}} = K_c G(s)$$

where
  K_c is the constant portion of the function and is termed the *steady state gain of the controller*.
  $G(s)$ is the dynamic portion
  (s) is usually omitted from the transfer function notation in the block diagram

Each block or element of the control system has its own characteristic transfer function.

An *overall transfer function* of a control system can be obtained by combining the transfer functions of the individual elements with the following rules:

1. For several transfer functions in series, the overall transfer function is the product of the individual transfer functions

2. In a single loop feedback system, the overall transfer function relating any two variables Y and X is given by the equation

$$\text{Overall transfer function} = \frac{Y}{X} = \frac{\pi_f}{1 \pm \pi_l}$$

where
  π_f = product of transfer functions between locations of X and Y
  π_l = product of transfer functions in the loop

Table 9.1 Transfer Functions of Control Actions

Controller Action	Functional Relationship between Controller Output and Error	Transfer Function
Proportional	$P = K_c\, e + b$	$\dfrac{P}{E} = K_c$
Integral	$P = \dfrac{1}{\tau_i} \int_0^t e\, dt$	$\dfrac{P}{E} = K_c\left(\dfrac{1}{\tau_i s}\right)$
Proportional + integral	$p = K_c\left(e + \dfrac{1}{\tau_i}\int_0^t e\, dt\right)$	$\dfrac{P}{E} = K_c\left(1 + \dfrac{1}{\tau_i s}\right)$
Proportional + derivative	$p = K_c\left(e + \tau_D \dfrac{de}{dt}\right)$	$\dfrac{P}{E} = K_c(1 + \tau_D s)$
Proportional + integral + derivative	$p = K_c\left(e + \dfrac{1}{\tau_i}\int_0^t e\, dt + \tau_D \dfrac{de}{dt}\right)$	$\dfrac{P}{E} = K_c\left(1 + \dfrac{1}{\tau_i s} + \tau_D s\right)$

Where p = fractional change in controller output, e = fractional change in error, K_c = controller gain, τ = integral time, and τ_D = derivative time.

A plus sign is to be used in the denominator when the feedback is negative and a minus sign when the feedback is positive. For an example of derivation of the differential equation and its Laplace transform, refer to Problem 9.1. Transfer functions for various control actions are given in Table 9.1.

PROPORTIONAL BAND

The basic relation for a proportional controller is

$$p = K_c\, e + b$$

where
 b = bias, which is usually adjusted to place the valve in its 50% open position with zero error.

The output of the controller equals the bias when there is no error.

$$\% \text{ Proportional Band} = \frac{100}{\text{Controller Gain}}$$

$$\text{or} \quad \text{Controller Gain} = \frac{100}{\% \text{ Proportional Band}}$$

Proportional band is defined as the percentage of the maximum range of the input variable required to cause 100 % change in the controller output.

$$\% \text{Proportional Band} = 100\, \frac{\Delta e}{\Delta e_{\max}}$$

where
 $\Delta e_{\max}$ is the maximum range of the error signal
 Δe is the change in error that gives a maximum change in the output of the controller

Substituting the value of K_c into the basic equation for the proportional controller, the following equation results

$$p = \frac{100}{\% \text{ Proportional Band}} e + 50\%$$

The proportional band is also called the proportional bandwidth and is often expressed as the percentage of bandwidth. The amount of the offset can be calculated using the following equation

$$\Delta e = \frac{\% \text{ Proportional Band}}{100} \Delta M$$

where
Δe = change in offset
ΔM change in measurement required by load upset

For a proportional pneumatic controller, the maximum output is 3 to 15 psig (in SI system, 20 to 100 kPa) or ΔP_{max} = 12 psi (in SI system, 80 kPa), which is related to error signal by the equation

$$\Delta P_{max} = K_c \Delta e$$

Combining the definition of % proportional band with the relation for ΔP_{max} gives

$$\% \text{ Proportional Band} = \frac{100}{K_c} \frac{\Delta P_{max}}{\Delta e_{max}}$$

Control action can be expressed by proportional bandwidth w. This is the error needed to cause a 100% change in controller output and is usually expressed in terms of chart width. The bandwidth w is given by

$$w = \frac{1}{k_c} \times 100$$

Some pneumatic controllers are calibrated in sensitivity units. With 4-inch chart width and 3 to 15 psig output range, the sensitivity of the controller is related to gain by

$$S = 3K_c \text{ psi/in travel}$$

TRANSIENT RESPONSE OF CONTROL SYSTEMS

The transient response of a control system is obtained with the application of some specified inputs called forcing functions. More commonly used forcing functions with their Laplace transforms are given in Table 9.2.

Table 9.2 Forcing Functions and Their Transforms

Input	Function, $f(t)$	Laplace Transform
Step	$aU(t)$	a/s
Ramp	at	a/s^2
Impulse	$a[\delta(t)]$	a
Sinusoidal	$\sin \omega t$	$\omega/(s^2 + \omega^2)$

An important consideration in the closed loop control system is the stability of the system. If the response function $y(t)$ in the time domain is bounded as $t \to \infty$ for all bounded inputs, the system is said to be stable; if $y(t) \to \infty$, the system is said to be unstable.

FIRST AND SECOND ORDER SYSTEMS

If the transfer function of a control system has a first order denominator, then the system is said to be of first order. Only one parameter, the time constant τ, describes the dynamic behavior of such a system. A second order system requires two parameters to describe its behavior. The denominator of the transfer function of a second order system is called *characteristic equation*. This can be expressed in the quadratic form as

$$\tau^2 s^2 + 2\xi\tau s + 1 = 0 \quad \text{or} \quad s^2 + 2\xi\omega s + \omega^2 = 0$$

where
ξ = damping coefficient
ω = radian frequency of oscillation = $\sqrt{1-\xi^2}/\tau$
τ = period of oscillation

The roots of the preceding equation will be real or complex depending upon the value of the damping coefficient ξ. The nature of the roots will determine the response function. For three possible types of roots, the system response will be as given in Table 9.3.

Of the three responses of the second order system listed in Table 9.3, the oscillatory or underdamped response is the most frequent in control systems. Hence, a number of special terms are in use to describe the underdamped response. These terms and their equations, which can be derived from the time response equation, are given below:

Overshoot for a unit step = $\exp\left(-\pi\xi/\sqrt{1-\xi^2}\right)$.

Decay ratio = $\exp\left(-2\pi\xi/\sqrt{1-\xi^2}\right)$ = (overshoot)^2.

Rise time, t_r: time required for the response to first reach its ultimate value.

Response time: time required for the response to come within ± certain percent (usually ±5%) of its ultimate value.

Period of oscillation: The radian frequency (rad/time) of the oscillations of an underdamped response is the coefficient of t in the sine term in the solution of the characteristic equation for this case. Thus

$$\text{radian frequency,}\ \omega = \frac{\sqrt{1-\xi^2}}{\tau}\ \text{radians/time}$$

Table 9.3 Responses of a Second Order System

ξ	Nature of Roots	Response
< 1	Complex	Oscillatory or underdamped
= 1	Real and equal	Critically damped
> 1	Real	Nonoscillatory or overdamped

The radian frequency ω and cyclical frequency f are related by $\omega = 2\pi f$. Hence, the cyclical frequency is given by

$$\text{Cyclical frequency} = f = \frac{1}{T} = \frac{1}{2\pi}\frac{\sqrt{1-\xi^2}}{\tau} \text{ also } T = \text{period of oscillation} =$$

time/cycle, from which, $T = \dfrac{2\pi\tau}{\sqrt{1-\xi^2}}$

Natural period of oscillation: When damping is eliminated ($\xi = 0$), the system oscillates continuously without attenuation in amplitude. The radian frequency under these undamped conditions is $1/\tau$ and is referred to as natural frequency ω_n.

$$\omega_n = 1/\tau$$

Natural frequency $= f_n = 1/T_n = 1/2\pi\tau$

$$\frac{\text{Actual frequency}}{\text{Natural frequency}} = \frac{f}{f_n} = \sqrt{1-\xi^2}$$

For more advanced topics, such as stability of higher order (>2) systems, Routh's test, frequency analysis, tuning of controllers, bode diagram, root locus method, and feed-forward control, the *AIChE. Modular Instruction Series* "Process Control" should be consulted.

REFERENCES

1. Das, D.K., and R. K. Prabhudesai, *Chemical Engineering:License Review*, 2nd ed., Kaplan AEC Education, 2004.
2. Perry, R.H., and D. Green, *Handbook for Chemical Engineers*, Platinum ed., New York, McGraw-Hill, 1999.
3. *AIChE Modular Instruction Series*, Process Control, Vol. 1–4.
4. Wylie, R., *Advanced Engineering Mathematics*, 4th ed., New York, McGraw-Hill.

PROBLEMS

9.1 Derive a transfer function for the transient variation in the liquid height for the tank filling system shown in Exhibit 9.1a and prepare a block diagram for the control system.

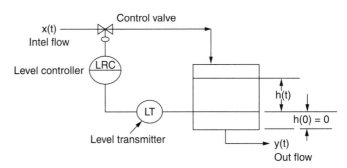

Exhibit 9.1a

9.2 A block diagram of a control system is shown in Exhibit 9.2. Obtain the overall transfer function $C(s)/R(s)$.

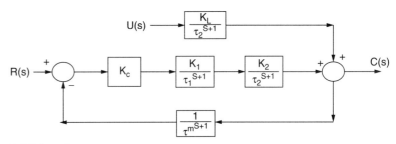

Exhibit 9.2

9.3 The liquid level in the bottom of a distillation column is controlled with a pneumatic proportional controller by throttling a control valve located in the bottom discharge line. The level may vary from 15 to 105 cm from the bottom tangent line of the column. With the controller set point held constant, the output pressure of the controller varies from 100 kPag (valve fully closed) to 20 kPag (valve fully open) as the level increases from 30 to 90 cm. from the bottom tangent line.
   (a) Find the % proportional band and the gain of the controller.
   (b) If the proportional band is changed to 80%, find the gain and the change in level required to cause the valve to go from fully open to a fully closed position.

SOLUTIONS

9.1 A material balance, over a difference time element Δt with assumption of constant density ρ, can be written as follows

$$\rho A h_2 - \rho A h_1 = \rho \overline{x(t)} \Delta t - \overline{\rho y(t)} \Delta t$$

which can be written in difference form after cancellation of ρ as follows

$$A \frac{\Delta h(t)}{\Delta t} = \overline{x(t)} - \overline{y(t)}$$

Taking the limit as $t \to 0$ gives

$$A \frac{dh(t)}{dt} = x(t) - y(t)$$

which on rearrangement becomes

$$x(t) = y(t) + A \frac{dh(t)}{dt}$$

where A is the cross sectional area of the tank. Laplace transformation gives

$$X(s) = Y(s) + A[sH(s) - h(0)]$$

For simplification, assume $h(0) = 0$. Therefore

$$X(s) = Y(s) + AsH(s)$$

Then the transfer function for the level system is given by

$$\frac{\text{Laplace transform of output}}{\text{Laplace transform of input}} = \frac{H(s)}{X(s) - Y(s)} = \frac{1}{As} = K_L G_L(s)$$

$1/As$ is the transfer function of the filling system. It comprises a constant portion $K_L = 1/A$ and $G_L(s) = 1/s$, which is the dynamic portion. Exhibit 9.1b shows the block diagram for the liquid filling system. The transfer function for the filling system is shown as $K_L G_L$

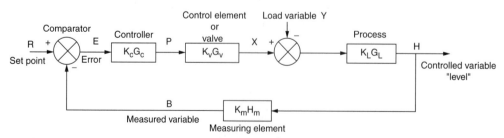

Exhibit 9.1b Block diagram for liquid filling system

9.2 For the transfer function between C and R

$$\pi_f = K_c \frac{K_1}{\tau_1 s+1} \frac{K_2}{\tau_2 s+1}$$

$$\pi_l = K_c \left[\frac{K_1}{\tau_1 s+1}\right]\left[\frac{K_2}{\tau_2 s+1}\right]\left[\frac{1}{\tau_m s+1}\right]$$

Then $\dfrac{C}{R} = \dfrac{K_c[K_1/(\tau_1 s+1)][K_2/(\tau_2 s+1)]}{1 + K_c K_1 K_2/[(\tau_1 s+1)(\tau_2 s+1)(\tau_m s+1)]}$

which simplifies to $\dfrac{C}{R} = \dfrac{K_c K_1 K_2 (\tau_m s+1)}{(\tau_1 s+1)(\tau_2 s+1)(\tau_m s+1) + K_c K_1 K_2}$

9.3 (a) Calculation of the proportional band

$$\text{Proportional band} = 100 \frac{\Delta e}{\Delta e_{\max}}$$

$$= \frac{\text{span of control variable for fully closed position of control valve}}{\text{full span or range}}$$

$$= \frac{0.9 - 0.3}{1.05 - 0.15} \times 100 = 67\%$$

$$\text{Gain of the controller} = \frac{\Delta P}{\Delta e} = \frac{100 - 20}{0.9 - 0.3} = 133.3 \text{ kPa/m}$$

(b) If the proportional band is 80%, $\dfrac{\Delta e}{\Delta e_{\max}} = 0.8(1.05 - 0.15) = 0.72 \text{ m}$

$$\text{Gain of controller,} \quad K_c = \frac{\Delta p}{\Delta e} = \frac{100 - 20}{0.72} = 11 \text{ kPa/m}$$

Process Equipment Design

OUTLINE

FLUID HANDLING EQUIPMENT 151
Centrifugal Pumps ■ Agitator

HEAT TRANSFER EQUIPMENT 156

PRESSURE VESSELS 156
Maximum Allowable Working Pressure ■ Design Pressure

MASS TRANSFER COLUMNS 157

REFERENCES 158

Process design involves application of the principles of engineering to design equipment that may be used to hold and process materials to make usable products. Design procedure for some major process equipment is given in this chapter.

FLUID HANDLING EQUIPMENT

In this section we will review design parameters for centrifugal pumps and agitators.

Centrifugal Pumps

In a centrifugal pump, the relation between volumetric flow rate, Q, mass flow rate, W, and specific gravity, s, is

$$Q = W/(500.4s) \tag{10.1}$$

The *net positive suction head available* ($NPSH_A$) is the total suction head in feet of liquid in absolute that is available in excess of the liquid vapor pressure, also expressed in feet of the same liquid, at the pump suction flange. Sufficient $NPSH_A$ is required to move the liquid into the eye of the impeller of a centrifugal pump without causing vaporization. Insufficient $NPSH_A$ causes the liquid to vaporize at the underside of the impeller vane near the base followed by the collapse of the vapor bubbles at high pressure beyond the tip of the impeller vane. This in turn results in pitting, vibration, noise, and reduction of discharge pressure, collectively known as cavitation. Similar conditions may occur in rotary and reciprocating pumps as well. Required NPSH is published by the manufacturers of the pump. It is a good engineering practice to have the available NPSH higher than required NPSH by a couple of feet.

$$\text{NPSH}_A = (P_1 - P_v - \Delta P_{f1})2.31/s + Z_1 + u^2/(2g_c) \quad (10.2a)$$

When NPSH is determined from the gage reading at the pump suction

$$\text{NPSH}_A = (P_1 \pm R_G - P_v - \Delta P_{f1})2.31/s + u^2/(2g_c) + Z_{1p} \quad (10.2b)$$

NPSH_A must always be positive, and greater than $\text{NPSH}_{required}$.

Suction Pressure, P_s, is given as

$$P_s = P_1 + 0.433Z_1 s - \Delta p_{f1} \quad (10.3)$$

Discharge pressure, P_d, is given as

$$P_d = P_2 + \Delta P_{f2} + 0.433Z_2 s \quad (10.4)$$

The relationship between total dynamic head (TDH) and discharge pressure, suction pressure, specific gravity, suction vessel pressure, destination vessel pressure, frictional drops, and elevation change is given as

$$\begin{aligned} TDH &= (P_d - P_s)2.31/s \\ &= (P_2 - P_1 + \Delta P_{f1} + \Delta P_{f2})2.31/s + Z_2 - Z_1 \end{aligned} \quad (10.5)$$

The equation for hydraulic horsepower, hhp, is

$$hhp = \frac{W(TDH)}{1.98 \times 10^6} = \frac{Q(TDH)s}{3960} = \frac{Q(P_d - P_s)}{1714} \quad (10.6)$$

The equation for brake horsepower, bhp, is

$$bhp = hhp/\varepsilon \quad (10.7)$$

Motor horsepower, mhp, is given as

$$mhp = bhp/\varepsilon_m \quad (10.8)$$

Maximum temperature rise due to pumping, assuming all power input goes to heat the fluid, ΔT, is given as

$$\Delta T = TDH/(778 C_p \varepsilon_{overall}) \quad (10.9a)$$
$$dT/d\theta = 5.1(bhp_{\text{shut-off}})/(VsC_p) \quad (10.9b)$$

When a running pump is put on a standby mode, such as in a batch operation, a minimum flow, Q_{min}, is to be drawn from the pump back to the suction vessel, normally through a restriction orifice, for thermal protection of the pump. The equation for minimum recirculation flow is

$$Q_{min} = \frac{5.1(bhp_{\text{shut-off}})}{s[C_p \cdot \Delta T_{safe} + 0.001285(TDH_{\text{shut-off}})]} \quad (10.10)$$

ΔT_{safe} = Saturation temperature corresponding to $[P_v + (0.433s)NPSH]$ − operating temperature, °F

This minimum flow should be added to the maximum demand to estimate the design capacity of a batch pump expected to run on a standby mode.

Another minimum flow, called the mechanical minimum flow, should be maintained to avoid excessive vibration and shaft deflection due to unbalanced radial loads. This flow is supplied by the manufacturer of the pump. The final minimum flow should be the greater of the two.

Pump affinity laws are summarized in the following equations:

	Within The Same Pump		Geometrically Similar Pumps
	D Constant	N Constant	N Constant
1. Capacity	$Q_2/Q_1 = N_2/N_1$	$Q_2/Q_1 = D_2/D_1$	$Q_2/Q_1 = (D_2/D_1)^3$ (10.11)
2. Head	$H_2/H_1 = (N_2/N_1)^2$	$H_2/H_1 = (D_2/D_1)^2$	$H_2/H_1 = (D_2/D_1)^2$ (10.12)
3. Power	$bhp_2/bhp_1 = (N_2/N_1)^3$	$bhp_2/bhp_1 = (D_2/D_1)^3$	$bhp_2/bhp_1 = (D_2/D_1)^5$ (10.13)

The same affinity laws apply to centrifugal fans and blowers. For design information for fans and compressors see Das and Prabhudesai.

Calculation of absolute pressure in process equipment requires the knowledge of absolute barometric pressure at the location under consideration. For air, the relationship between the local barometric pressure at an elevation less than 11 km from sea level and standard barometric pressure at sea level is given by Darby as follows:

$$P_{local} = 14.696 \left[1 - \frac{0.0065 \Delta Z}{288} \right]^{5.26433} \quad (10.14)$$

where
- bhp = brake horsepower
- $bhp_{shut-off}$ = brake horsepower at shut-off
- C_p = heat capacity of fluid pumped, Btu/lb·°F
- D = impeller diameter, ft
- $dT/d\theta$ = rate of temperature rise in a dead-headed pump, °F/minute
- g_c = Newton's law proportionality factor, 32.17 lb·ft/s²·lbf
- H = total dynamic head, ft
- hhp = hydraulic horsepower
- mhp = motor horsepower
- N = impeller speed, revolutions per minute (rpm)
- $NPSH_A$ = net positive suction head available, ft
- P_1 = pressure in suction vessel, psia
- P_2 = pressure in discharge vessel, psia
- P_{local} = local barometric pressure, psia
- ΔP_{f1} = total frictional drop in suction line, psi
- ΔP_{f2} = total frictional drop in discharge line, psi
- P_d = pump discharge pressure, psia
- P_s = pump suction pressure, psia
- P_v = vapor pressure of fluid, psia, at the operating temperature

Q = volumetric flow rate, gpm
Q_{min} = minimum flow rate of pump for thermal protection, gpm
R_G = gage reading, psig; use minus sign for vacuum gage
s = specific gravity of fluid being pumped, dimensionless
TDH = Total dynamic head, ft
$TDH_{shut\text{-}off}$ = Total dynamic head at shut-off, ft
ΔT = Maximum temperature rise due to pumping, °F
u = Velocity of fluid in suction line, ft/s
V = holding capacity of the casing of the pump, gallon
W = mass flow rate, lb/h
Z_1 = vertical distance between the liquid surface in the suction vessel and the centerline of the pump, ft; a negative value of Z_1 must be used in Equations (10.2a), (10.3), and (10.5) if the liquid surface is below the centerline of the pump
Z_{1p} = distance between the centerline of the gage and the center line of the pump, ft; use minus sign if the centerline of the gage is below the centerline of the pump
Z_2 = vertical distance between the liquid surface in the discharge vessel for subsurface discharge or the point of discharge for above-surface discharge and the centerline of the pump, ft
ΔZ = change in elevation from sea level, m
ε = pump efficiency, decimal; it is the ratio of hydraulic horsepower to the motor horsepower, known as brake horsepower, delivered to the pump
ε_m = motor efficiency, decimal; it is the ratio of the horsepower supplied by a motor to the pump, known as brake horsepower, to the nameplate horsepower of the motor
$\varepsilon_{overall} = \varepsilon \varepsilon_m$

Agitator

The agitator behaves like a pump. The primary pumping capacity, Q, is given by

$$Q = 4.33 \times 10^{-3} N_Q N D^3 \qquad (10.14)$$

The apparent superficial velocity, v_b, due to pumping effect of the impeller is given by

$$v_b = Q/(A \times 7.48) \qquad (10.15)$$

Chemscale, N_I, is an arbitrary scale of agitation (1 = mild, 10 = violent).

$$N_I = v_b/6 \qquad (10.16)$$

where v_b, expressed in ft/minute, varies from 6 (mild) to 60 (violent)

Tip speed is the linear velocity of the tip of the impeller blade.

$$TS = 0.2618 DN \qquad (10.17)$$

There is an arbitrary scale of agitation on the basis of tip speed: low ($TS = 200 - 500$), medium ($TS = 500 - 900$), high ($900 - 2000$).

Turnover time is the time to circulate the entire contents of the vessel once and is given as

$$\theta_Q = V/Q \tag{10.18}$$

Mixing time (Khang and Levenspiel), θ, is the time to mix materials to satisfy a specified fluctuation of concentration of a key ingredient (for example, ±0.1%).

For conventional axial flow impellers (Reynolds number $> 1 \times 10^5$)

$$(N/K)(D/T)^2 = N_P = 0.9 \tag{10.19a}$$

$$a = 2e^{-K\theta} \tag{10.19b}$$

$$\theta = 0.9[\ln(2/a)](T/D)^2/N \tag{10.19c}$$

With $a = 0.001$

$$\theta = 6.84(T/D)^2/N \tag{10.19d}$$

The equation for Reynolds number, N_{Re}, is

$$N_{Re} = 10.7sND^2/\mu \tag{10.20}$$

Power number, N_P, is

$$N_P = \frac{1.523 \times 10^{13} P}{sN^3 D^5} = \frac{P_I}{sN_{rps}^3 D_I^5} \tag{10.21}$$

Torque, τ, is given as

$$\tau = 63025 P/N \tag{10.22}$$

The equation for motor load (three-phase power), is

$$HP = \frac{1.73 \, (\text{Voltage})(\text{ampere})(\text{motor efficiency, fraction})(\text{power factor})}{746} \tag{10.23}$$

where
- a = amplitude of concentration variation, decimal
- A = vessel cross section, ft^2
- D = impeller diameter, inch
- D_I = impeller diameter, m
- K = amplitude decay constant, min^{-1}
- N = rpm (revolution per minute) of impeller
- N_{rps} = revolution per second of impeller
- N_I = chemscale
- N_Q = pumping number (dimensionless). It depends on the type of impeller and impeller Reynolds number, and is given by the manufacturer of agitator.
- N_P = power number, dimensionless, dependent on impeller design and Reynolds number.
- P = agitator power, HP
- P_I = agitator power, kW
- Q = primary pumping capacity, gpm
- s = specific gravity of fluid, dimensionless

T = Tank diameter, inch
TS = tip speed, ft/minute
v_b = apparent superficial velocity, ft/min
V = working capacity of the vessel, gallon
θ = mixing time, minute when N is rpm; second, when N_I in rps in equation (10.19c & d)
θ_Q = turnover time, minute
τ = torque, inch-lb
μ = viscosity, cP

HEAT TRANSFER EQUIPMENT

Heat transfer equipment is common equipment for process industries. It may be shell-and-tube type, plate-type, or block-type equipment. The theoretical basis of heat transfer is provided in Chapter 7.

PRESSURE VESSELS

The design and construction of pressure vessels are governed by the codes of the American Society of Mechanical Engineers (ASME) and the standards of the American Petroleum Institute (API). Three sections of ASME that govern the pressure vessels are

(1) ASME Section I for fired pressure vessels such as boilers.

(2) ASME Section VIII Division 1 and 2 for unfired metallic pressure vessels. Section VIII Division I is used to design most of the pressure vessels required by the chemical industries. It employs relatively approximate formulas to specify a vessel whose design pressure (internal or external) is greater than 15 psig but does not exceed 3000 psig with an internal diameter exceeding 6 inches. Division 2 uses more complex and rigorous mathematical analysis and is used for applications involving severe service (such as toxic chemicals), cyclic operations, design pressure exceeding 3000 psig, and so on.

(3) ASME Section X is used for fiber-reinforced thermosetting plastic pressure vessels for internal design pressure exceeding 15 psig but not exceeding 3000 psig, temperature range of −65 °F to 250 °F for non-lethal service. External pressure must not exceed 15 psig. For example, for a jacketed vessel, external design pressure may easily exceed 15 psig; therefore, coded plastic vessels are not generally jacketed.

API standards (12D, 12F, 620, 650, 650F) cover design of pressure vessels with design pressure not exceeding 15 psig.

The following sections elaborate on aspects of internal pressure per ASME Section VIII Division 1.

Maximum Allowable Working Pressure

Maximum allowable working pressure (MAWP) is the least of the pressure ratings of the various components of a pressure vessel. The pressure ratings of the components are back-calculated by using the code formulae from known plate thickness after allowances are made for tolerances, corrosion, thinning due to fabrication, and other loadings. The MAWP at the hot and corroded condition is valid only when the vessel is hydrotested at 1.5 times the MAWP at new and cold condition (because of higher thickness) and stamped along with the design temperature on a plate attached to the vessel.

Design Pressure

This pressure, taken at the highest connection of the pressure vessel, is added to the hydrostatic head, and the resulting pressure is used to calculate the theoretical thickness of a plate.

Table 10.1 summarizes the formulas for calculating plate thickness and maximum allowable working pressure.

Table 10.1 Plate Thickness and Maximum Allowable Working Pressure

	Wall Thickness Inch	Maximum Allowable Working Pressure, psig	
Cylindrical shell	$t = \dfrac{PR}{SE - 0.6P}$	$P = \dfrac{SEt}{R + 0.6t}$	(10.24)
Sphere and hemispherical head	$t = \dfrac{PR}{2SE + 0.8P}$	$P = \dfrac{2SEt}{R - 0.8t}$	(10.25)
2:1 Ellipsoidal head	$t = \dfrac{PR}{SE + 0.9P}$	$P = \dfrac{SEt}{R - 0.9t}$	(10.26)
Conical head	$t = \dfrac{PR}{\cos\alpha(SE + 0.4P)}$	$P = \dfrac{SEt(\cos\alpha)}{R - 0.4t(\cos\alpha)}$	(10.27)
ASME flanged & dished head (torispherical) $(L/r = 16\ 2/3)$[1]	$t = \dfrac{0.885PL}{SE + 0.8P}$	$P = \dfrac{SEt}{0.885L - 0.8t}$	(10.28)

In Table 10.1
- E = joint efficiency, decimal, varies from 0.45 to 1.0
- L = outside radius of dish for ASME flanged and dished head, inch
- P = design pressure or maximum allowable working pressure, psig
- R = outside radius of vessel, inch
- r = knuckle radius of a flanged and dished head
- S = allowable stress of material corresponding to the design temperature, psi
- t = wall thickness, inch; corrosion allowance, if applicable, and other appropriate allowances are added to this thickness, and the nearest commercially available plate thickness is chosen
- α = one-half of the included apex angle(cone angle) of a conical head

Notes:
1. For $L/r < 16\ 2/3$, see ASME Section VIII
2. For external pressure, see ASME Section VIII

MASS TRANSFER COLUMNS

Mass transfer columns are broadly divided into packed columns and plate columns. Sizing of plate columns involves selection of tray spacing and sizing of the diameter and overall height. Plate spacing should be sufficient for the separation of the entrained liquid from the vapor before it reaches the next higher plate. Plate spacing is a function of many variables. It generally varies from 12 inches to 48 inches. Flooding limits for bubble-cap and perforated plates are plotted in Fig. 18-10 of Perry and Green. For a specified tray spacing, and known vapor and liquid flow rates, known physical properties, such as liquid density, vapor density, and liquid surface tension, the flooding velocity can be calculated. Design velocity ranging from 80% to 95% of the flooding velocity is used to estimate the net column area from a known vapor load. Before selecting the design velocity, check Fig. 18-22 of Perry and Green so that entrainment is less than 0.1 lb of liquid per pound of liquid flow. Depending on the chosen tray type, an area required for the downcomer is to be added to get the total cross section area of the column. Liquid loading in gpm and a recommended liquid velocity (gpm/ft^2) are used to estimate the downcomer area.

Determination of the actual number of plates (N_a) requires the knowledge of overall plate efficiency (E_{oc}) and the number of theoretical plates (N_t):

$$E_{oc} = \frac{N_t}{N_a}$$

The overall plate efficiency can be determined by empirical methods such as O'Connell correlation (Fig 18-23a in Perry and Green), Backowski correlation (Equation 18-29 in Perry and Green) or by the theoretical predictive method of AIChE (as shown in Perry and Green, p. 18-15).

REFERENCES

1. Das, D. K., and R. K. Prabhudesai, *Chemical Engineering License Review*, 2nd ed., Kaplan AEC Education, 2004.
2. *Chemical Engineering*, February 23, 1981, pp. 83–85.
3. Darby, R., *Chemical Engineering Fluid Mechanics*, 2nd ed., New York, Marcel Dekker, Inc, p. 91.
4. Khang, S. J., and Levenspiel, O., *Chemical Engineering Science*, 1976, vol. 31, pp. 5610–577, Pergamon Press.
5. Perry, R. H., and D. Green, *Chemical Engineers' Handbook*, 6th ed. New York, McGraw-Hill Book Co., 1984, p. 18-7.

PROBLEMS

10.1 A centrifugal pump equipped with a motor having 7.5 KW (brake) at shut-off head and holding 0.006 m^3 of fluid of specific gravity 1.1 and heat capacity of 4186 J/kgK in the casing is running dead-headed. Find the rate of temperature rise, assuming all brake horsepower is used to heat the trapped fluid.

10.2 Determine the minimum flow for thermal protection of a pump delivering water at 122°F with a shut-off head of 1032 ft and shut-off bhp of 48. The available NPSH is 12 ft.

10.3 A small amount of hydrochloric acid is added for pH control to a baffled reactor, 2.74 m inside diameter, fitted with a 1.92 m axial flow impeller. Estimate the rps and power of the agitator if the specific gravity is 1.1 and mixing time is 10 seconds with ±0.1% concentration fluctuation required to avoid side reactions. Assume a power number of 0.9.

10.4 Calculate the minimum thickness required for a cylindrical shell with outside radius of 54 inches. The shell is constructed of carbon steel with allowable stress of 17,500 psi. The joint efficiency is 85%, and the corrosion allowance is 0.125 inch. The design pressure is 100 psig, with no hydrostatic head.

10.5 Size the diameter of a column, using the Fair method for bubble-cap trays, at 80% of the flooding velocity with 24 inches of tray spacing.

Use the following correlation to calculate the vapor velocity through the net flow area (i.e., total column internal area minus the downcomer area):

$$U_{nf} = C_{sb,\text{flood}} \left(\frac{\sigma}{20}\right)^{0.2} \left(\frac{\rho_l - \rho_g}{\rho_g}\right)^{0.5}$$

where

    U_{nf} = vapor velocity through the net flow area at flooding condition
    σ = surface tension, dyne/cm
    ρ = density, lb/ft^3

Subscripts: l = liquid, g = gas or vapor.

To find the value of $C_{sb,\text{flood}}$, the capacity parameter, ft/s, use the following equation to find the flow parater F_{lv}:

$$F_{lv} = \frac{L}{G}\left(\frac{\rho_g}{\rho_l}\right)^{0.5}$$

where
    L/G = liquid-gas mass ratio, dimensionless, under consideration
    (ρ_g/ρ_l) = vapor density/liquid density, dimensionless

Use the following table to find $C_{sb,\text{flood}}$:

F_{lv}	0.1	0.2	0.3	0.5
$C_{sb,\text{flood}}$	0.33	0.28	0.24	0.18

Data available are the following:

Vapor flow: 300,000 lb/h
Liquid flow: 284,600 lb/h
Vapor density: 3 lb/ft^3
Liquid density: 30 lb/ft^3
Surface tension: 20 dyn/cm
Recommended downcomer velocity: 100 gpm/ft^2

The required diameter in an integer figure is
a. 4 ft   c. 12 ft
b. 16 ft   d. 9 ft

10.6 A pump has a suction lift of 12 ft (vertical lift from the liquid surface in an open container to the center line of the centrifugal pump). The liquid has a specific gravity of 1.1, and the local barometric pressure is 14.68 psia. The suction line friction loss is 2.5 psi and the suction line velocity is 2 ft/s. The vapor pressure of the liquid at the pumping temperature is 0.3 psia. The required NPSH is 10 ft of the liquid. Examine if the available NPSH is adequate at (1) the location of the specified barometric pressure and (2) at a location that is 5000 ft above sea level.

10.7 The following table shows the specification of the composition of the feed, distillate, and bottoms of a distillation column (the components are listed in order of decreasing K-values):

Component	Feed (mol%)	Distillate (mol%)	Bottoms (mol%)
Methane	25	41.74	
Ethane	8	13.36	
Propane	26	43.11	0.44
n-Butane	18	1.79	42.11
n-Pentane	12		29.92
n-Hexane	11		27.43
Total	100	100	100

Based on the preceding specification
(a) The light key is
   a. Methane   c. Propane
   b. Ethane   d. n-Butane

(b) The heavy key is
   a. Propane   c. n-Hexane
   b. n-Butane

10.8 In Problem 10.7, moles of distillate per 100 moles of feed are
   a. 100   c. 67.9
   b. 59.9   d. 20.7

10.9 In Problem 10.7, $(I_{LK/HK}) = 1.98$. The minimum number of theoretical stages including the reboiler as a stage is close to
   a. 12   c. 6
   b. 4   d. 18

   The minimum theoretical stages may be calculated from

   $$N_{min} = \frac{\log\left[\left(\frac{x_{LK}}{x_{HK}}\right)_D \left(\frac{x_{HK}}{x_{LK}}\right)_B\right]}{\log(\alpha_{LK/HK})_{av}}$$

10.10 In Problem 10.7, if the specified separation requires 14 theoretical stages, then counting from the top, the location of the feed tray is close to
   a. 6   c. 8
   b. 3   d. 9

   Use the Underwood-Fenske equation:

   $$\frac{N_R}{N_S} = \frac{\log\left[\left(\frac{x_D}{x_F}\right)_{LK} \left(\frac{x_F}{x_D}\right)_{HK}\right]}{\log\left[\left(\frac{x_F}{x_B}\right)_{LK} \left(\frac{x_B}{x_F}\right)_{HK}\right]}$$

   N = number of theoretical stages
   x = liquid mole fraction of a key component
   Subscripts: B = bottoms, D = distillate, HK = heavy key, LK = light key, R = rectification section, S = stripping section.

10.11 In Problem 10.7 the following data are available per 100 lbmoles of feed/h (enthalpies are defined in this problem as the heat content of stream with reference to a chosen datum temperature):

   Feed enthalpy = -4.772×10^6 Btu/h
   Distillate enthalpy = -2.348×10^6 Btu/h
   Bottoms enthalpy = -2.47×10^6 Btu/h
   External Reflux ratio = 5
   Latent heat of overhead vapor = 5556.5 Btu/lbmol

   The reboiler duty for the specified feed rate in million Btu/h is
   a. 1.997   c. 2.013
   b. 1.945   d. 2.105

SOLUTIONS

10.1 $0.006 \text{ m}^3 \times 1.1 \times 1000 \text{ kg/m}^3 \times 4186 \text{ J/kgK} \times dT/d\theta = 7.5 \times 10^3$
$$dT/d\theta = 0.2715 \text{ °C/s}$$

10.2 From Equation (10.10):

$$Q_{min} = \frac{5.1(bhp_{shut-off})}{s[C_p \cdot \Delta T_{safe} + 0.001285(TDH_{shut-off})]}$$

T_{safe} = Saturation temperature corresponding to $[P_v + (0.443s)NPSH]$ − operating temperature, °F

P_v of water at 122°F = 1.8 psia
$0.433sNPSH = 0.433 \times .99 \times 12$ = 5.1 psia
$\overline{}$ 6.9 psia

Saturation temperature of water corresponding to 6.9 psia = 176 °F

$$\Delta T_{safe} = 176 - 122 = 54$$

$$Q_{min} = \frac{5.1 \times 48}{0.99 \times (1 \times 54 + 0.001285 \times 1032)} = 4.5 \text{ gpm}$$

10.3 From Equation (10.19c)

$$\theta = 0.9\ln[(2/a)](T/D)^2/N$$
$$N = 0.9\ln[(2/a)](T/D)^2/\theta$$
$$a = 0.001, \ T/D = 2.74/1.92 = 1.427, \ \theta = 10 \text{ s}$$
$$N = 1.39 \text{ rps} \simeq 1.4 \text{ rps}$$

From Equation (10.21)

$$N_p = \frac{P_l}{N_{rps}^3 D_l^5 s} = 0.9$$

$$P_l = N_p N_{rps}^3 D_l^5 s$$
$$= 0.9(1.4)^3(1.92)^5 \times 1.1$$
$$= 70.9 \text{ KW}$$

10.4 $\quad t = \dfrac{PR}{SE - 0.4P} = \dfrac{100 \times 54.125}{17{,}500 \times .85 - 0.4 \times 100} = 0.365 \text{ inch}$

Theoretical minimum thickness = 0.365 + 0.125 = 0.49 inch. Note that corrosion allowance is added to the outer radius.

10.5 The vapor velocity through the net area of a distillation column may be computed by the given equation:

$$U_{nf} = C_{sb,flood}\left(\frac{\sigma}{20}\right)^{0.2}\left(\frac{\rho_l - \rho_g}{\rho_g}\right)^{0.5}$$

To calculate $C_{sb,flood}$, one has to calculate the flow parameter F_{lv}:

$$F_{lv} = \frac{L}{G}\left(\frac{\rho_g}{\rho_l}\right)^{0.5}$$
$$= (284,600/300,000)(3/30)^{0.5} = 0.3$$

From the given table $C_{sb,flood} = 0.24$.

The flooding velocity, U_{nf}:

$$U_{nf} = 0.24\left(\frac{30-3}{3}\right)^{0.5} = 0.72 \text{ ft/s}$$

The design velocity, $U_{nd} = 0.8 \times 0.72 = 0.576$ ft/s
Net flow area = $(300,000/3600 \times 3 \times 0.576) = 48.23$ ft^2
Liquid flow = $(284,600 \times 62.4/500.4 \times 30) = 1182.99$ gpm
Downcomer area = $1182.99/100 = 11.83$ ft^2
Total area = $48.23 + 11.83 = 60.06$ ft^2
Diameter = $(4 \times 60.06/3.1416)^{0.5} = 8.74$ ft

Answer: D

10.6 From Equation (10.2a):

$$NPSH_A = (P_1 - P_v - P_{f1})2.31/s + Z_1 + u^2/(2g_c)$$

(1) Location of barometric pressure 14.68 psia

Given:

P_1 = pressure in suction vessel = 14.68 psia, since container is open
P_v = vapor pressure of liquid = 0.3 psia
P_{f1} = total frictional drop in suction line = 2.5 psi
s = specific gravity of liquid = 1.1
$Z_1 = -12$ ft, negative since it is a lift, i.e., the liquid has to move up to the suction of the pump
u = fluid velocity = 2 ft/s

Unknown:

$NPSH_A$

Solution:

Substituting the above values:

$NPSH_A = (14.68 - 0.3 - 2.5)2.31/1.1 - 12 + 2^2/(2 \times 32.17)$
$= 13.01$ ft

Since the available NPSH is higher than required NPSH of 10 ft, it is adequate at above location.

(2) Location of elevation 5000 ft above sea level.

$$P_{local} = 14.696\left[1 - \frac{0.0065\Delta Z}{288}\right]^{5.26433}$$

$NPSH_A = (P_1 - P_v - P_{fl})2.31/s + Z_1 + u^2/(2g_c)$

Given:

Z = elevation from sea level = 5000 ft = 1524 m
P_v = vapor pressure of liquid = 0.3 psia
P_{fl} = total frictional drop in suction line = 2.5 psi
s = specific gravity of liquid = 1.1
Z_1 = −12 ft, negative since it is a lift, i.e., the liquid has to move up to the suction of the pump
u = fluid velocity = 2 ft/s

Unknown:

Local barometric pressure

$P_{local} = 14.696[1 - 0.0065 \times 1524/288]^{5.26433} = 12.22$ psia = P_1
$NPSH_A = (P_1 - P_v - P_{fl})2.31/s + Z_1 + u^2/(2g_c)$
$= (12.22 - 0.3 - 2.5)2.31/1.1 - 12 + 2^2/(2 \times 32.17)$
$= 7.844$ ft

Since this is lower than the required NPSH, the available NPSH is not adequate at the elevation of 5000 ft.

10.7 By definition, the key components are two components whose separation is specified in the problem statement.

(a) Light key is propane. Answer: C

(b) Heavy key is *n*-butane. Answer: B

10.8 Since all methane ends up in distillate

moles distillate (0.4174) = 100(0.25)
or, moles of distillate = 25/0.4174 = 59.9 moles. Answer: B

10.9 **Given:**

Distillate: $x_{LK} = 0.4311$, $x_{HK} = 0.0179$
Bottoms: $x_{LK} = 0.0044$, $x_{HK} = 0.4221$
$(I_{LK/HK})_{av} = 1.98$

Substituting the values in the given equation:

$$N_{min} = \frac{\log\left[\left(\frac{0.4311}{0.0179}\right)\left(\frac{0.4221}{0.0044}\right)\right]}{\log(1.98)} = 11.34$$

Answer: A

10.10 **Given:**

Underwood-Fenske equation

$$\frac{N_R}{N_S} = \frac{\log\left[\left(\frac{x_D}{x_F}\right)_{LK}\left(\frac{x_F}{x_D}\right)_{HK}\right]}{\log\left[\left(\frac{x_F}{x_B}\right)_{LK}\left(\frac{x_B}{x_F}\right)_{HK}\right]}$$

x_D for the light key = 0.4311
x_F for the light key = 0.26
x_F for the heavy key = 0.18
x_D for the heavy key = 0.0179
x_B for the light key = 0.0044
x_B for the heavy key = 0.4221
N_T = total theoretical stages = 14

Calculation:

Substituting appropriate values in the Underwood-Fenske equation:

$$\frac{N_R}{N_S} = \frac{\log\left[\left(\frac{0.4311}{0.26}\right)\left(\frac{0.18}{0.0179}\right)\right]}{\log\left[\left(\frac{0.26}{0.0044}\right)\left(\frac{0.4221}{0.18}\right)\right]} = 0.5706$$

$$\frac{N_R + N_S}{N_S} = \frac{N_T}{N_S} = 1.5706, \quad \text{where } N_T = \text{total theoretical stages.}$$

$$N_S = \frac{N_T}{1.5706} = \frac{14}{1.5706} = 8.91$$

$$N_R = 14 - 8.91 = 5.09$$

Answer: (A)

10.11 **Given:**

Feed rate (F) = 100 lbmol/h
x_F = 0.26 based on light key
x_D = 0.4311
x_B = 0.0044
Feed enthalpy (Fh_F) = -4.772×10^6 Btu/h
Distillate enthalpy (Dh_D) = -2.348×10^6 Btu/h
Bottoms enthalpy (Bh_B) = -2.47×10^6 Btu/h
External Reflux ratio (R) = 5
Latent heat of overhead vapor (λ) = 5556.5 Btu/lbmol

Assumption:

No heat loss from column and other equipment wall and associated piping.

Calculation:

Overall energy balance with distillate and bottoms as the only outlet process streams:

$$F(h_F) + Q_B = D(h_D) + B(h_B) + Q_C \quad (1)$$
$$Q_C = D(R + 1)\lambda \quad (2)$$

Similarly, overall material and component balances give:

$$F = D + B \quad (3)$$
$$F(x_F) = D(x_D) + B(x_B) \quad (4)$$

Where
  F = feed rate, lbmol/h
  D = distillate rate, lbmol/h
  B = bottoms rate, lbmol/h
  Q_B = reboiler heat load, Btu/h
  Q_C = condenser cooling load, Btu/h

Substituting appropriate given values in equations 3 and 4:

$$D = 59.9 \text{ lbmol/h}$$

Substituting the calculated value of D and given values of R and λ in equation 2:

$$Q_C = 1.997(10^6) \text{ Btu/h}$$

Substituting the calculated value of Q_C and enthalpies in million Btu/h

$$-4.772 + Q_B = -2.348 - 2.476 + 1.997$$

or, $Q_B = 1.945$ million Btu/h

Answer: B

CHAPTER 11

Computer and Numerical Methods

OUTLINE

INTRODUCTION 168
Number Systems ■ Data Storage ■ Data Transmission ■ Programming ■ Algorithmic Flowcharts ■ Pseudocode

SPREADSHEETS 173
Relational References ■ Arithmetic Order of Operation ■ Absolute References

NUMERICAL METHODS 174
Root Extraction ■ Newton's Method

NUMERICAL INTEGRATION 176
Euler's Method ■ Trapezoidal Rule

NUMERICAL SOLUTIONS OF DIFFERENTIAL EQUATIONS 177
Reduction of Differential Equation Order

PACKAGED PROGRAMS 178

GLOSSARY OF COMPUTER TERMS 178

INTRODUCTION

Number Systems

The number system most familiar to everyone is the decimal system based on the symbols 0 through 9. This base-10 system requires 10 different digits to create the representation of numbers.

A far simpler system is the binary number system. This base-2 system uses only the characters 0 and 1 to represent any number. A binary representation 110, for example, corresponds to the number.

$$1 \times 2^2 + 1 \times 2^1 + 0 \times 2^0 = 4 + 2 + 0 = 6$$

Similarly, the binary number 1010 would be

$$1 \times 2^3 + 0 \times 2^2 + 1 \times 2^1 + 0 \times 2^0 = 8 + 0 + 2 + 0 = 10$$

The digital computer is based on the binary system of on/off, yes/no, or 1/0. For this reason it may at times be necessary to convert a decimal number (like 12) to a binary number (12 would be 1100).

Example 11.1

The binary number 1110 corresponds to what decimal (base 10) number?

Solution

$$1 \times 2^3 + 1 \times 2^2 + 1 \times 2^1 + 0 \times 2^0 = 8 + 4 + 2 = 14$$

The conversion of a decimal number to a binary number can be achieved by the method of remainders as follows. A decimal integer is divided by 2, giving an integer quotient and a remainder. This process is repeated until the quotient becomes 0. The remainders (in the reverse order) form the binary number. The following example illustrates this process.

Example 11.2

Convert decimal number 43 to a binary number.

Solution

	Quotient		Remainder
43 ÷ 2 =	21	+	1
21 ÷ 2 =	10	+	1
10 ÷ 2 =	5	+	0
5 ÷ 2 =	2	+	1
2 ÷ 2 =	1	+	0
1 ÷ 2 =	0	+	1

Answer: $(43)_{10} = (101011)_2$

Conversion of a decimal fraction to a binary fraction is accomplished by successive multiplication by 2. The integer portion of the number after multiplication is the binary digit. The fractional part is repeatedly multiplied until it becomes 0. The integers in the correct order form the binary fraction.

A large number of binary digits is difficult for humans to use. Consequently, for human-machine interaction, other powers of base-2 number systems are used. Octal (base-8) and hexadecimal (base-16) number systems are generally used: the latter is the most common. The hexadecimal system is a shorthand method of representing the value of four binary digits at a time. Since the hexadecimal system requires 16 different characters to represent decimal digits 0 through 15, the letters A, B, C, D, E, and F are used to represent decimal numbers 10 through 15, respectively.

The conversion from binary to hexadecimal is accomplished by grouping binary digits (bits) into groups of four bits starting from the binary point and proceeding to the left and to the right. Each group of four bits is then converted to the corresponding hexadecimal digit. Conversion from decimal to hexadecimal may be carried out in a manner similar to that for decimal to binary conversion, with the divisor 2 (or multiplier in the case of fractions) replaced by 16.

Example 11.3

Convert the base 10 integer 458 to base 16 equivalent value.

	Quotient		Remainder	Hexadecimal Digit
458 ÷ 16	28	+	10	A
28 ÷ 16	1	+	12	C
1 ÷ 16	0	+	1	1

Data Storage

Memory hardware in a modern computer is semiconductor based. Types of memory include random access memory (RAM), read-only memory (ROM), programmable read-only memory (PROM), and erasable programmable read-only memory (EPROM). The amount of RAM (main memory) in a microcomputer is typically 256 MB to 1 GB and is volatile.

For permanent mass data storage, magnetic and optical disk drive units are available and generally included. Magnetic disk drives (hard drives) are made up of several platters each with one or more read/write heads, depending on the density and size of storage and on data access speed. Platters turn at high speed (3000 to 7200 rpm).

Data on a platter are organized into tracks and sectors. Tracks are the concentric storage areas, and sectors are pie-shaped subdivisions of each track. The data are also organized in cylinders, which are the same numbered tracks on all drive platters.

The hard drives are generally fixed, although a number of portable hard drives are now available that connect to a microcomputer (PC) through its universal serial bus (USB) or parallel port. The storage capacity of these fixed drives ranges from 10 GB at the low end to over 240 GB. Portable drives typically have 40-GB storage capacities. Optical disk drives can be read-only (R/O). WORM drives (write once, read many) can be written to by the user. Other drives, such as CD and DVD drives, can be read-only (CD-ROM and DVD-ROM) or read and write (CD-RW and DVD-RW). These types of storage provide a convenient way of transferring large amounts of data between computers. Several other devices, such as memory cards, memory sticks, thumb drives, flash drives, Zip disks, and floppy disks, serve the same purpose of porting data between PCs and in some cases between other digital devices (such as still and video cameras and printers), although with smaller storage capacity.

The storage capacity of a diskette depends on its size, the recording density, number of tracks, and number of sides. Although at one time larger size (8-inch and $5\frac{1}{4}$-inch) floppy diskettes were used, the $3\frac{1}{2}$-inch rigid diskette is the current standard.

In addition to the storage capacity of these fixed disk drives, several other parameters may be used to describe their performance:

- *Average seek time:* Average time it takes to move a head from one location to a new location.

- *Track-to-track seek time:* The time required to move a head from one track to an adjacent one.

- *Latency or rotational delay:* The time it takes for a head to access a particular sector to read or write. On the average, this is the time required for one-half revolution of the disk.

- *Average access time:* The time required to move to a new sector and read the data. This is the sum of latency and average seek time.

Data Transmission

A computer generates digital (pulse) signals (represented as on/off or 0/1), but often these signals cannot be transmitted on telephone lines over long distances. These signals are converted to analog form (usually tones) and then modulated and transmitted over analog transmission lines. The transmitted signals at the receiving end are demodulated (converted to tone signals) and then converted to digital pulses. The device designed to do this conversion and communication is called a **modem** (derived from modulation-demodulation). Typical modems available for a PC are capable of operating at up to 56 kilobits per second for transmission over voice grade telephone lines. Digital Subscriber Line (DSL) and broadband cable and satellite service use modems that are capable of operating at much higher speeds, usually in megabits per second (Mbps). For transmission and reception of data, modems are required at both ends that are capable of communicating with each other using appropriate protocol.

There are two common methods of communication: asynchronous and synchronous. In an asynchronous system each character is preceded and followed by start and stop bits. Thus, every 8-bit character requires 10 bits for transmission. There is a 20% penalty in transmission rate in overheads due to the required start and stop bits. Asynchronous communication is, however, the most common method in data communication at low data rates.

In a synchronous system the data is transmitted continuously. Start and stop bits are not needed, but synchronous communication requires clock synchronization. This is accomplished by sending special characters for synchronization. Separation of a bit stream into individual characters is accomplished by counting bits from the start of the previous character. Synchronous transmission is approximately 20% faster than asynchronous and is frequently used in large-volume, high-speed data transmission. Characters are sent in blocks and special (synchronization) characters are placed at the beginning of the block and within it to ensure that the receiving clock remains accurate (synchronized). Thus, there is an overhead penalty in this case as well, but it is far less than 20%.

In any transmission system, errors are bound to occur. The methods of ensuring accuracy of transmission and reception are called communication protocols and transmission standards. Frequently, error checking and correcting technique is employed for accuracy. One or more parity bits may be added as a means of checking and correcting. Another method is for both receiver and sender to calculate a block check digit derived from each block of characters sent.

Programming

Computer programming may be thought of as a four-step process:

1. Defining the problem
2. Planning the solution
3. Preparing the program
4. Testing and documenting the program

Once the problem has been carefully defined, the basic programming work of planning the computer solution and preparing the detailed program can proceed. In this section the discussion will be limited to two ways to plan a computer program to solve a problem: algorithmic flowcharts and pseudocode.

Algorithmic Flowcharts

An algorithmic flowchart is a pictorial representation of the step-by-step solution of a problem using standard symbols. Some of the commonly used shapes are shown in Fig. 11.1. Consider the following simple problem.

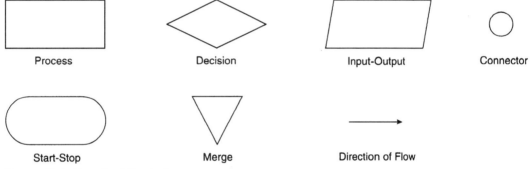

Figure 11.1 Flowchart Symbols

Example 11.4

A present sum of money (P) at an annual interest rate (I), if kept in a bank for N years, would amount to a future sum (F) at the end of that time according to the equation $F = P(1 + I)^N$. Prepare a flowchart for $P = \$100$, $I = 0.07$, and $N = 5$ years. Then compute and output the values of F for all values of N from 1 to 5. Exhibit 1 shows a flowchart for this situation.

Example 11.5

Consider the flowchart in Exhibit 2.
The computation does which of the following?

(a) Inputs hours worked and hourly pay and outputs the weekly paycheck for 40 hours or less.

(b) Inputs hours worked and hourly pay and outputs the weekly paycheck for hours worked including over 40 hours at premium pay.

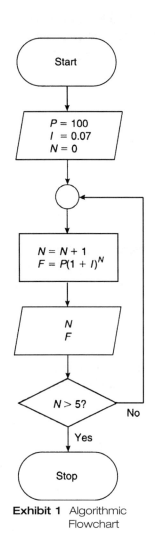

Exhibit 1 Algorithmic Flowchart

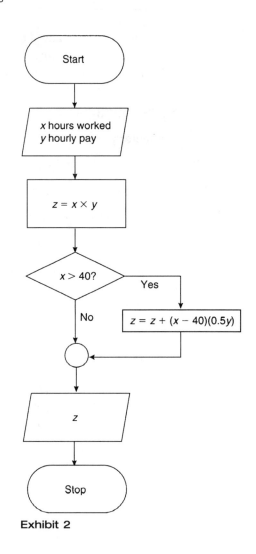

Exhibit 2

Solution

The answer is (b).

Pseudocode

Pseudocode is an English-like-language representation of computer programming. It is carefully organized to be more precise than a simple statement, but may lack the detailed precision of a flowchart or of the computer program itself.

Example 11.6

Prepare pseudocode for the computer problem described in Example 11.4.

Solution

```
INPUT P, I, and N = 0
DOWHILE N < 5
COMPUTE N = N + 1
    F = P (1 + I)^N
OUTPUT N, F
IF N > 5 THEN ENDDO
```

SPREADSHEETS

For today's engineers the ability to create and use spreadsheets is essential. The most popular spreadsheet programs are Microsoft's **Excel**, Novell's **Quattro Pro**, and Lotus **1-2-3**. Each of these programs uses similar construction, methods, operators, and relative references.

Three types of information may be entered into a spreadsheet: text, values, and formulas. Text includes labels, headings, and explanatory text. Values are numbers, times, or dates. Formulas combine operators and values in an algebraic expression.

A **cell** is the intercept of a column and a row. Its location is based upon its column-row location; for example, B3 is the intercept of column B and row 3. Column labels are across the top of the spreadsheet and row labels are on the side. To change a cell entry, the cell must be highlighted using either an address or pointer.

A group of cells may be called out by using a **range**. Cells A1, A2, A3, A4 could be called out using the range reference A1:A4 (or A1..A4). Similarly, the range A2, B2, C2, D2 could use the range reference A2:D2 (or A2..D2).

In order to call out a block of cells, a range callout might be A2:C4 (or A2..C4) and would reference the following cells:

A2 B2 C2

A3 B3 C3

A4 B4 C4

Formulas may include cell references, operators (such as +,–,*,/), and functions (such as SUM, AVG). The formula SUM(A2:A6) or SUM(A2..A6) would be evaluated as equal to A2 + A3 + A4 + A5 + A6.

Relational References

Most spreadsheet references are relative to the cell's position. For example, if the content of cell A5 contains B4, then the value of A5 is the value of the cell up one and over one. The relational reference is most frequently used in tabulations, as in the following example for an inventory where cost times quantity equals value and the sum of the values yields the total inventory cost.

Inventory	Valuation		
	A	B	C
1 Item	Cost	Quantity	Value
2 box	5.2	2	10.4
3 tie	3.4	3	10.2
4 shoe	2.4	2	4.8
5 hat	1.0	1	1.0
6 Sum			26.4

In C2, the formula is A2*B2, in C3 A3*B3, and so forth. For the summation the function SUM is used; for example, SUM(C2:C5) in C6.

Instead of typing in each cell's formula, the formula can be copied from the first cell to all of the subsequent cells by first highlighting cell C2 then dragging the mouse to include cell C5. The first active cell, C2, would be displayed in the edit window. Typing the formula for C2 as = A2*B2 and holding the control key down and pressing the enter key will copy the relational formula to each of the highlighted cells. Since the call is relational, the formula in cell C3 is evaluated as A3*B3. In C4 the cell is evaluated A4*B4. Similarly, edit operations to include copy or fill operations simplify the duplication of relational formula from previously filled-in cells.

Arithmetic Order of Operation

Operations in equations use the following sequence for precedence: exponentiation, multiplication or division, followed by addition or subtraction. Parentheses in formulas override normal operator order.

Absolute References

Sometimes one must use a reference to a cell that should not be changed, such as a data variable. An absolute reference can be specified by inserting a dollar sign ($) before the column-row reference. If B2 is the data entry cell, then by using B2 as its reference in another cell, the call will always be evaluated to cell B2. Mixed reference can be made by using the dollar sign for only one of the elements of the reference. For example, the reference B$2 is a mixed reference in that the row does not change but the column remains a relational reference.

The power of spreadsheets makes repetitive calculations very easy. In many operations, periods of time are normally new columns. Each element can become a row item with changes in time becoming the columns.

All spreadsheet programs allow for changes in the appearance of the spreadsheet. Headings, borders, or type fonts are usual customizing tools.

NUMERICAL METHODS

This portion of numerical methods includes techniques of finding roots of polynomials by the Routh-Hurwitz criterion and Newton methods, Euler's techniques of numerical integration and the trapezoidal methods, and techniques of numerical solutions of differential equations.

Root Extraction

Routh-Hurwitz Method (without Actual Numerical Results)

Root extraction, even for simple roots (i.e., without imaginary parts), can become quite tedious. Before attempting to find roots, one should first ascertain whether they are really needed or whether just knowing the area of location of these roots will suffice. If all that is needed is knowing whether the roots are all in the left half-plane of the variable (such as is in the s-plane when using Laplace transforms—as is frequently the case in determining system stability in control systems), then one may use the Routh-Hurwitz criterion. This method is fast and easy even for higher-ordered equations. As an example, consider the following polynomial:

$$p_n(x) = \prod_{m=1}^{n}(x - x_m) = x^n + a_1 x^{n-1} + a_2 x^{n-2} + \cdots + a_{n-1} \quad (11.1)$$

Here, finding the roots, x_m, for $n > 3$ can become quite tedious without a computer; however, if one only needs to know if any of the roots have positive real parts, one can use the Routh-Hurwitz method. Here, an array is formed listing the coefficients of every other term starting with the highest power, n, on a line, followed by a line listing the coefficients of the terms left out of the first row. Following rows are constructed using Routh-Hurwitz techniques, and after completion of the array, one merely checks to see if all the signs are the same (unless there is a zero coefficient—then something else needs to be done) in the first column; if none, no roots will exist in the right half-plane. In case of zero coefficient, a simple technique is used; for details, see almost any text dealing with stability of control systems. A short example follows.

$F(s) = s^3 + 3s^2 + 2s + 10$ Array:	s^3	1	2	Where the s^1 term is formed as
	s^2	3	10	$(3 \times 2 - 10 \times 1)/3 = -\frac{4}{3}$. For details,
$= (s + ?)(s + ?)(s + ?)$	s^1	$-\frac{4}{3}$	0	refer to any text on control
	s^0	10	0	systems or numerical methods.

Here, there are two sign changes: one from 3 to $-4/3$, and one from $-4/3$ to 10. This means there will be two roots in the right half-plane of the s-plane, which yield an unstable system. This technique represents a great savings in time without having to factor the polynomial.

Newton's Method

The use of Newton's method of solving a polynomial and the use of iterative methods can greatly simplify a problem. This method utilizes synthetic division and is based upon the remainder theorem. This synthetic division requires estimating a root at the start, and, of course, the best estimate is the actual root. The root is the correct one when the remainder is zero. (There are several ways of estimating this root, including a slight modification of the Routh-Hurwitz criterion.)

If a $P_n(x)$ polynomial (see Eq. (11.1)) is divided by an estimated factor $(x - x_1)$, the result is a reduced polynomial of degree $n-1$, $Q_{n-1}(x)$, plus a constant remainder of b_{n-1}. Thus, another way of describing Eq. (11.1) is

$$P_n(x)/(x - x_1) = Q_{n-1}(x) + b_{n-1}/(x - x_1) \quad \text{or} \quad P_n(x) = (x - x_1)Q_{n-1}(x) + b_{n-1} \quad (11.2)$$

If one lets $x = x_1$, Eq. (11.2) becomes

$$P_n(x = x_1) = (0)Q_{n-1}(x) + b_{n-1} = b_{n-1} \quad (11.3)$$

Equation (11.3) leads directly to the remainder theorem: "The remainder on division by $(x - x_1)$ is the value of the polynomial at $x = x_1$, $P_n(x_1)$."*

---

* Gerald & Wheatley, *Applied Numerical Analysis*, 3rd ed., Addison-Wesley, 1985.

Newton's method (actually, the Newton-Raphson method) for finding the roots for an nth-order polynomial is an iterative process involving obtaining an estimated value of a root (leading to a simple computer program). The key to the process is getting the first estimate of a possible root. Without getting too involved, recall that the coefficient of x^{n-1} represents the sum of all of the roots and the last term represents the product of all n roots; then the first estimate can be "guessed" within a reasonable magnitude. After a first root is chosen, find the rate of change of the polynomial at the chosen value of the root to get the next, closer value of the root x_{n+1}. Thus the new root estimate is based on the last value chosen:

$$x_{n+1} = x_n - P_n(x_n)/P'_n(x_n),$$
where $P'_n(x_n) = dP_n(x)/dx$ evaluated at $x = x_n$ (11.4)

NUMERICAL INTEGRATION

Numerical integration routines are extremely useful in almost all simulation-type programs, design of digital filters, theory of z-transforms, and almost any problem solution involving differential equations. And since digital computers have essentially replaced analog computers (which were almost true integration devices), the techniques of approximating integration are well developed. Several of the techniques are briefly reviewed below.

Euler's Method

For a simple first-order differential equation, say $dx/dt + ax = af$, one could write the solution as a continuous integral or as an interval type one:

$$x(t) = \int^t [-ax(\tau) + af(\tau)]d\tau \tag{11.5a}$$

$$x(kT) = \int^{kT-T} [-ax + af]d\tau + \int_{kT-T}^{kT} [-ax + af]d\tau = x(kT - T) + A_{\text{rect}} \tag{11.5b}$$

Here, A_{rec} is the area of $(-ax + af)$ over the interval $(kT - T) < \tau < kT$. One now has a choice looking back over the rectangular area or looking forward. The rectangular width is, of course, T. For the forward-looking case, a first approximation for x_1 is**

$$x_1(kT) = x_1(kT - T) + T[ax_1(kT - T) + af(kT - T)]$$
$$= (1 - aT)x_1(kT - T) + aTf(kT - T) \tag{11.5c}$$

Or, in general, for Euler's forward rectangle method, the integral may be approximated in its simplest form (using the notation $t_{k+1} - t_k$ for the width, instead of T, which is $kT-T$) as

$$\int_{t_k}^{t_{k+1}} x(\tau)d\tau \approx (t_{k+1} - t_k)x(t_k) \tag{11.6}$$

** This method is as presented in Franklin & Powell, *Digital Control of Dynamic Systems*, Addison-Wesley, 1980, page 55.

Trapezoidal Rule

This trapezoidal rule is based upon a straight-line approximation between the values of a function, $f(t)$, at t_0 and t_1. To find the area under the function, say a curve, is to evaluate the integral of the function between point a and b. The interval between these points is subdivided into subintervals; the area of each subinterval is approximated by a trapezoid between the end points. It will be necessary only to sum these individual trapezoids to get the whole area; by making the intervals all the same size, the solution will be simpler. For each interval of delta t (i.e., $t_{k+1} - t_k$), the area is then given by

$$\int_{t_k}^{t_{k+1}} x(\tau)\, d\tau \approx (1/2)(t_{k+1} - t_k)[x(t_{k+1}) + x(t_k)] \tag{11.7}$$

This equation gives good results if the delta t's are small, but it is for only one interval and is called the "local error." This error may be shown to be $-(1/12)(\text{delta } t)^3 f''(t = \xi_1)$, where ξ_1 is between t_0 and t_1. For a larger "global error" it may be shown that

$$\text{Global error} = -(1/12)(\text{delta } t)^3 \, [f''(\xi_1) + f''(\xi_2) + \cdots + f''(\xi_n)] \tag{11.8}$$

Following through on Eq. (11.8) allows one to predict the error for the trapezoidal integration. This technique is beyond the scope of this review or probably the examination; however, for those interested, please refer to pages 249–250 of the previously mentioned reference to Gerald & Wheatley.

NUMERICAL SOLUTIONS OF DIFFERENTIAL EQUATIONS

This solution will be based upon first-order ordinary differential equations. However, the method may be extended to higher-ordered equations by converting them to a matrix of first-ordered ones.

Integration routines produce values of system variables at specific points in time and update this information at each interval of delta time as T (delta $t = T = t_{k+1} - t_k$). Instead of a continuous function of time, $x(t)$, the variable x will be represented with discrete values such that $x(t)$ is represented by $x_0, x_1, x_2, \ldots, x_n$. Consider a simple differential equation as before as, based upon Euler's method,

$$dx/dt + ax = f(t).$$

Now assume the delta time periods, T, are fixed (not all routines use fixed step sizes); then one writes the continous equations as a difference equation where $dx/dt \approx (x_{k+1} - x_k)/T = -ax_k + f_k$ or, solving for the updated value, x_{k+1},

$$x_{k+1} = x_k - Tax_k + Tf_k \tag{11.9a}$$

For fixed increments by knowing the first value of $x_{k=0}$ (or the initial condition), one may calculate the solution for as many "next values" of x_{k+1} as desired for some value of T. The difference equation may be programmed in almost any high-level language on a digital computer; however, T must be small as compared to the shortest time constant of the equation (here, $1/a$).

The following equation—with the "f" term meaning "a function of" rather than as a "forcing function" term as used in Eq. (11.5a)—is a more general form of Eq. (11.9a). This equation is obtained by letting the notation (x_{k+1}) become $y[k+1\,\Delta t]$ and is written (perhaps somewhat more confusingly) as

$$y[(k+1)\Delta t] = y(k\Delta t) + \Delta t f[y(k\Delta t), k\Delta t]\tag{11.9b}$$

Reduction of Differential Equation Order

To reduce the order of a linear time-dependent differential equation, the following technique is used. For example, assume a second-order equation: $x'' + ax' + bx = f(t)$. If we define $x = x_1$ and $x' = x_1' = x_2$, then

$$x_2' + ax_2 + bx_1 = f(t)$$
$$x_1' = x_2 \text{ (by definition)}$$
$$x_2' = -b_{x1} - ax_2 + f(t)$$

This technique can be extended to higher-order systems and, of course, be put into a matrix form (called the state variable form). And it can easily be set up as a matrix of first-order difference equations for solving digitally.

PACKAGED PROGRAMS

Most currently available packaged simulation programs use algorithms not necessarily based upon Euler's methods but more advanced methods such as the Runge-Kutta method. Automatic variable step size methods like Milne's may also be used. However, as mentioned before, these routines are all built into the packaged programs and may be transparent to the user. The user of a specialized program may be without knowledge of the high-level language being employed (except for certain modifications).

GLOSSARY OF COMPUTER TERMS

Accumulators	Registers that hold data, addresses, or instructions for further manipulations in the ALU
Address bus	Two-way parallel path that connects processors and memory containing addresses
AI	Artificial intelligence
Algorithm	A sequence of steps applied to a given data set that solves the intended problem
Alphanumeric data	Data containing the characters a, b, c, ..., z, 0, 1, 2, ..., 9
ALU	Arithmetic and logic unit
ASCII	American Standard Code for Information Interchange, 7 bit/character (Pronounced AS-key)
Asynchronous	Form of communications in which message data transfer is not synchronous, with the basic transfer rate requiring start/stop protocol
Baud rate	Bits per second
BIOS	Basic input/output system
Bit	0 or 1
Buffer	Temporary storage device
Byte	8 bits
Cache memory	Fast look-ahead memory connecting processors with memory, offering faster access to often-used data
Channel	Logic path for signals or data
CISC	Complex instruction-set control

Glossary of Computer Terms

Clock rate	Cycles per second
Control bus	Separate physical path for control and status information
Control unit	Fetches, decodes instructions to control the operations of registers and ALUs
CPU	Central Processing Unit, the primary processor
Data buffer	Temporary storage of data
Data bus	Separate physical path dedicated for data
Digital	Discrete level or valued quantification, as opposed to analog or continuous valued
Duplex communication	Communications mode where data is transmitted in both directions at the same time
Dynamic memory	Storage that must be continually hardware-refreshed to retain valid information
EBCDIC	Extended Binary Coded Decimal Interchange Code—8 bits/character (pronounced EB-see-dick)
EPROM	Erasable programmable read-only memory
Expert systems	Programs with AI which learn rules from external stimuli
Floppy disk	Removable disk media in various sizes, $5\frac{1}{4}''$, $3\frac{1}{2}''$
Flowchart	Graphical depiction of logic using shapes and lines
Gbyte (GB)	Gigabytes: 1,073,741,824 or 2^{30} bytes
Half-duplex communication	Two-way communications path in which only one direction operates at a time (transmit or receive)
Handshaking	Communications protocol to start/stop data transfer
Hard disk	Disk that has nonremovable media
Hardware	Physical elements of a system
Hexadecimal	Numbering system (base 16) that uses 0–9, A, B, …, F
Hierarchical database	Database organization containing hierarchy of indexes/keys to records
I/O	Input/output devices such as terminal, keyboard, mouse, printer
IR	Instruction register
Kbytes (KB)	Kilobytes: 1024 or 2^{10} bytes
LAN	Local area network
LIFO	Last in–first out
LSI	Large scale integration
Main memory	That memory seen by the CPU
Mbytes (MB)	Megabytes: 1,048,576 or 2^{20} bytes
Memory	Generic term for random access storage
Microprocessor	Computer architecture with Central Processing Unit in one LSI chip
MODEM	Modulator-demodulator
MOS	Metal oxide semiconductor
Multiplexer	Device that switches several input sources, one at a time, to an output
Nibble	Four bits
Nonvolatile memory	As opposed to volatile memory, does not need power to retain its present state
Number systems	Method of representing computer data as human-readable information
OCR	Optical character recognition
OS	Operating system
OS memory	Memory dedicated to the OS, not usable for other functions
Parallel interface	A character (8-bit) or word (16-bit) interface with as many wires as bits in interface plus data clock wire.
Parity	Method for detecting errors in data: one extra bit carried with data, to make the sum of one bit in a data stream even or odd
PC	Program counter, or personal computer
Peripheral devices	Input/output devices not contained in main processing hardware
Program	A sequence of computer instructions
PROM	Programmable read-only memory
Protocols	Established set of handshaking rules enabling communications
Pseudocode	An English-like way of representing structured programming control structures

RAM	Random access memory
Real time/Batch	Method of program execution: real-time implies immediate execution; batch mode is postponed until run on a group of related activities
Relational database	Database organization that relates individual elements to each other without fixed hierarchical relationships
RISC	Reduced instruction set computer
ROM	Read-only memory
Scratchpad memory	High-speed memory, in either hardware or software
Sequential storage	Memory (usually tape) accessed only in sequential order ($n, n + 1, \ldots$)
Serial interface	Single data stream that encodes data by individual bit per clock period
Simplex communication	One-way communication
Software	Programmable logic
Stacks	Hardware memory organization implementing LIFO access
Static memory	Memory that does not require intermediate refresh cycles to retain state
Structured programming	Use of programming constructs such as Do-While or If-Then, Else
Synchronous	Communications mode in which data and clock are at same rate
Transmission speed	Rate at which data is moved, in baud (bits per second, bps)
Virtual memory	Addressable memory outside physical address bus limits through use of memory mapped pages
Volatile memory	Memory whose contents are lost when power is removed
VRAM	Video memory
Words	8, 16, or 32 bits
WYSIWYG	What you see is what you get
16-bit	Basic organization of data with 2 bytes per word
32-bit	Basic organization of data with 4 bytes per word
64-bit	Basic organization of data with 8 bytes per word
80386	Intel's microprocessor architecture based on 16–data address bus, extended virtual memory, external math coprocessor
80486	Upgrade to 80386 incorporating math coprocessor within VLSI
80586	Intel's microprocessor architecture based on 16-bit data, 32-bit address bus

PROBLEMS

11.1 In spreadsheets, what is the easier way to write B1 + B2 + B3 + B4 + B5?
   a. Sum (B1:B5)
   b. (B1..B5) Sum
   c. @B1..B5SUM
   d. @SUMB2..B5

11.2 The address of the cell located at row 23 and column C is
   a. 23C
   b. C23
   c. C.23
   d. 23.C

11.3 Which of the following is **not** correct?
   a. A CD-ROM may not be written to by a PC.
   b. Data stored in a batch processing mode is always up-to-date.
   c. The time needed to access data on a disk drive is the sum of the seek time, the head switch time, the rotational delay, and the data transfer time.
   d. The methods for storing files of data in secondary storage are sequential file organization, direct file organization, and indexed file organization.

11.4 Which of the following is false?
   a. Flowcharts use symbols to represent input/output, decision branches, process statements, and other operations.
   b. Pseudocode is an English-like description of a program.
   c. Pseudocode uses symbols to represent steps in a program.
   d. Structured programming breaks a program into logical steps or calls to subprograms.

11.5 In pseudocode using DOWHILE, the following is true:
   a. DOWHILE is normally used for decision branching.
   b. The DOWHILE test condition must be false to continue the loop.
   c. The DOWHILE test condition tests at the beginning of the loop.
   d. The DOWHILE test condition tests at the end of the loop.

11.6 A spreadsheet contains the following formulas in the cells:

	A	B	C
1		A1 +1	B1 +1
2	A1 ^2	B1^2	C1^2
3	Sum (A1:A2)	Sum (B1:B2)	Sum (C1:C2)

If 2 is placed in cell A1, what is the value in cell C3?
   a. 12
   b. 20
   c. 8
   d. 28

11.7 A matrix contains the following:

	A	B	C	D
1		3	4	5
2	2	A$2		
3	4			
4	6			

If you copy the formula from B2 into D4, what is the equivalent formula in D4?
a. A$2
b. C4
c. C4
d. C$2

11.8 A processing system is processor limited when sorting on small tables in memory. Which of the following would speed up computations?
a. Adding more main memory
b. Adding virtual memory
c. Adding cache memory
d. Adding peripheral memory

11.9 A small PC processing system is performing large (1-MB) matrix operations that are currently I/O limited since memory is limited to 1MB. Which of the following would speed up computations?
a. Adding more main memory
b. Adding virtual memory
c. Adding cache memory
d. Adding peripheral memory

11.10 Transmission Protocol: Serial, asynchronous, 8-bit ASCII, 1 Start, 1 Stop, 1 Parity bit, 9600 bps. How long will it take for a 1-Kbyte file to be transmitted through the link?
a. 0.85 s
b. 0.96 s
c. 1.07 s
d. 1.17 s

11.11 Transmission Protocol: Serial, synchronous, 8-bit ASCII, 1 Parity bit, 9600 bps. How long will it take for a 1-Kbyte file to be transmitted through the link?
a. 0.85 s
b. 0.96 s
c. 1.07 s
d. 1.17 s

11.12 The hexadecimal number 2DB.A is most nearly equivalent to which decimal number?
a. 731.625
b. 731.10
c. 453.625
d. 341.10

11.13 The decimal number 1938.25 is most nearly equivalent to which hexadecimal number?
a. $(792.25)_{16}$
b. $(792.4)_{16}$
c. $(279.4)_{16}$
d. $(279.04)_{16}$

11.14

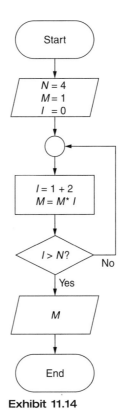

Exhibit 11.14

The output value of M in Exhibit 11.14 is closest to
a. 1   c. 8
b. 2   d. 48

11.15 Pseudocode can best be described as
   a. A simple letter-substitution method of encryption.
   b. An English-like language representation of computer programming.
   c. A relational operator in a database.
   d. The way data are stored on a diskette.

11.16 The pseudocode that best represents Exhibit 11.16 is:

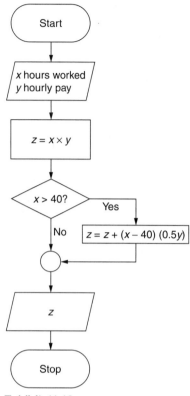

Exhibit 11.16

a. INPUT X,Y
    WAGE = X × Y
    IF X less than 40 THEN
    WAGE = WAGE + OVERTIME PAY
    END IF
    OUTPUT WAGE

b. INPUT hours worked and hourly pay
    WAGE = hours worked × hourly pay
    IF hours worked greater than 40 THEN
    WAGE = WAGE + (hours worked-40)(overtime wage supplement)
    END IF
    OUTPUT WAGE

c. INPUT hours worked and hourly pay
    WAGE = (hours worked − 40)(overtime rate) + 40(hourly pay)
    OUTPUT WAGE

d. PRINT hours worked and hourly pay
    WAG = hours worked × hourly pay
    IF hours worked equals 40 THEN
    WAGE = hours worked × hourly pay + (hours worked − 40) × overtime pay
    END IF
    OUTPUT WAGE

SOLUTIONS

11.1 A. Sum (B1:B5) or @Sum (B1..B5).

11.2 B.

11.3 B. Batch mode processing always has delays in updating the database.

11.4 C. Pseudocode does not use symbols but uses English-like statements such as IF-THEN and DOWHILE.

11.5 C. IF-THEN is normally used for branching. The DOWHILE test condition must be true to continue branching, and the test is done at the beginning of the loop. The DOUNTIL test is done at the end of the loop.

11.6 B. Plugging 2 into cell A1 of the spreadsheet produces the following matrix:

	A	B	C
1	2	3	4
2	4	9	16
3	6	12	20

The value of C3 is 20.

11.7 D. The formula contains mixed references. The $ implies absolute row reference, whereas the column is relative. The result of any copy would eliminate any answer except for the absolute row 2 entry. The relative column reference A gets replaced by C. The cell contains C$2.

11.8 C. Processor-limited sorting on small tables suggests either speeding up processor cycles or providing faster memory. Since speeding up clock is not an option, making memory faster is the answer. Adding more memory or virtual memory, however, does nothing for small tables. Only adding cache memory would allow the CPU to fetch recently used data without full memory cycles, thereby speeding up the sorting process.

11.9 A. Processing is I/O limited because not all of the matrix can fit into memory. Since this is a large matrix, cache memory probably would not affect processing. The best solution is adding more main memory to fit this problem entirely in memory.

11.10 D.

$$1 \text{ Kbyte} = 2^{10} \text{ bytes} = 1024$$

$$1024 \text{ bytes} + 3 \text{ overhead bits/byte} = 1024 \text{ bytes } (8 \text{ bits/byte} + 3 \text{ bits/byte overhead})$$
$$= 11{,}264 \text{ bits}$$

$$\text{Minimum transmission time} = 11{,}264 \text{ bits}/9600 \text{ bps} = 1.17 \text{ s.}$$

11.11 **B.**

$$1 \text{ Kbyte} = 2^{10} \text{ bytes} = 1024$$
$$1024 \text{ bytes} + 1 \text{ overhead bit/byte} = 1024 \text{ bytes } (8 \text{ bits/byte} + 1 \text{ bit/byte overhead})$$
$$= 9216 \text{ bits}$$
$$\text{Minimum transmission time} = 9216 \text{ bits}/9600 \text{ bps} = 0.96 \text{ s}$$

11.12 **A.**

$$(2DB)_{16} = 2 \times 16^2 + 13 \times 16^1 + 11 \times 16^0 = 512 + 208 + 11 = 731$$
$$(.A)_{16} = 10 \times 16^{-1} = .625$$
$$(2DB.A)_{16} = 731.625$$

11.13 **B.**

	Quotient		Remainder	Hexadecimal Digit
1938 ÷ 16 =	121	+	2	2
121 ÷ 16 =	7	+	9	9
7 ÷ 16 =	0	+	7	7

$$(1938)_{10} = (792)_{16}$$

	Integer		Fraction	Hexadecimal Digit
0.25 × 16 =	4	+	0.00	4

$$(.25)_{10} = (.4)_{16}$$
$$(1938.25)_{10} = (792.4)_{16}$$

11.14 **D.**

11.15 **B.**

11.16 **B.**

Process Safety

OUTLINE

THRESHOLD LIMIT VALUES 187
Time-Weighted Average for Different Exposures to Same Toxicant ■ Exposure to More Than One Toxicant ■ Mixture of Liquids ■ Required Dilution Air

FIRE AND EXPLOSION ISSUES 189
Fire Tetrahedron ■ Dust Explosion Pentagon

CLASSIFICATION OF LIQUIDS FOR THERMAL HAZARDS ANALYSIS 191

COMPUTATION OF FLASH POINT 192

CALCULATING FLAMMABILITY LIMITS 192

STOICHIOMETRY OF COMBUSTION REACTION AND ESTIMATION OF FLAMMABILITY LIMITS 193

INERTING AND PURGING 194
Sweep Through Purging ■ Pressure Purging and Vacuum Purging

LIMITING AND EXCESS REACTANT 195

REFERENCES 195

Process engineers are responsible not only for the most economical design and operation of processes but also for the safe design and operation of equipment. This chapter deals with the fundamentals of process safe.

THRESHOLD LIMIT VALUES

The threshold limit values (TLV) are the highest concentration of a toxicant in air that can be tolerated for a given length of time without any adverse effect. There are five types of TLV as shown next. Of these, the first three have been established by the American Conference of Governmental Industrial Hygienists (ACGIH):

1. *TLV-TWA*: The time weighted average of a toxicant for a normal eight-hour work day or 40-hour work week to which a worker may be repeatedly exposed,

day after day, without adverse effect. Excursions above the limit are allowed if compensated by excursions below the limit.

2. *TLV-STEL*: The short term exposure limit is the maximum concentration to which a person may be exposed continuously up to 15 minutes without adverse effect. Such effects include intolerable irritation, chronic tissue change, narcosis to increase accident proneness, impairment of self-rescue, and material reduction of worker efficiency. No more than four excursions are permitted per day with at least 60 minutes between exposure periods, provided that the daily TLV-TWA is not exceeded.

3. *TLV-C*: The ceiling limit is the concentration that should not be exceeded any time.

4. *PEL*: The permissible exposure limit is the maximum concentration exposure allowed by OSHA 29CFR 1910.1000 as the eight-hour time-weighted average. OSHA also has an excursion limit.

5. *IDLH*: The concentration of the toxicant that is immediately dangerous to life or health is the maximum concentration from which one could escape within 30 minutes without escape-impairing symptoms or any irreversible health effects.

TLV are expressed either in parts per million (ppm) by volume or milligram per cubic meter (mg/m^3). The relation between ppm by volume and mg/m^3 is as follows.

$$1 \text{ ppm} = \frac{PM}{0.08205T} \text{ mg/m}^3$$

where
- P = pressure in atm
- T = temperature in K
- M = molecular weight

Time-Weighted Average for Different Exposures to Same Toxicant

If a worker is exposed to different concentrations of the same toxicant at different lengths of time, the time-weighted average of concentration is calculated as follows.

$$TWA = \frac{C_1 t_1 + C_2 t_2 + \cdots + C_n t_n}{t_1 + t_2 + \cdots + t_n}$$

where C_i is the concentration (ppm) to which the worker is exposed for length of time t_i

Exposure to More Than One Toxicant

The effect of more than one toxicant with different TLV-TWA may be determined by evaluating the following value:

$$\text{Value} = \frac{C_1}{TLV_1} + \frac{C_2}{TLV_2} + \cdots$$

where C_1 is the concentration in ppm of the toxicant whose threshold limit value is TLV_1 (ppm) and so on

If the value of the preceding summation exceeds 1, then the exposure limit is exceeded.

Mixture of Liquids

To determine TLV in a vapor space of a mixture of liquids of known vapor pressure and composition, the equations are

$$y_i = \frac{P_i^0 x_i}{\sum P_i^0 x_i}$$

$$TLV_{\text{mixture}} = \frac{1}{\sum \frac{y_i}{TLV_i}}$$

$$TLV_i \text{mixture} = y_i (TLV)_{\text{mixture}}$$

where

TLV_i = threshold limit value (ppm) of pure component i (when present by itself)
P_i^0 = pure component vapor pressure of component i at the mixture temperature
TLV_{mixture} = threshold limit value of the vapor mixture of toxicants
TLV_i mixture = TLV of component i in the mixture
x_i = mole fraction of component i in the mixture of liquid
y_i = mole fraction of component i in the vapor phase

Required Dilution Air

The requirement of dilution air to maintain toxicant concentration below the TLV is given by

$$\text{Dilution air (CFM)} = \frac{(3.87 \times 10^8) W}{M(TLV) K}$$

where
CFM = cubic feet per minute
W = flow rate or generation rate of toxicant, lb/min
M = molecular mass of toxicant
TLV = threshold limit value of toxicant, ppm
K = nonideal mixing factor
= 1 for perfect mixing
= 0.1 to 0.5 for most real situation

FIRE AND EXPLOSION ISSUES

Fire and explosion prevention are key aspects of process safety.

Fire Tetrahedron

A fire is caused by the presence of sufficient quantity of each of the following three sources and the fourth criteria:

1. Fuel: such as gasoline, hydrogen, wood, etc.
2. Oxidizer: such as oxygen, chlorine, ammonium nitrite, etc.
3. Ignition source: such as heat, mechanical impact, mechanical or electrical sparks or electrostatic discharge, etc.
4. Chemical chain reaction necessary to propagate fire

Dust Explosion Pentagon

In addition to three sources mentioned previously, two additional factors contribute to dust explosion:

1. Confinement: this can be safeguarded by explosion venting
2. Dispersion: such as distribution of dusts in air

Terminology

Autoignition Temperature (AIT): The temperature above which a flammable substance ignites itself by drawing sufficient amount of oxidizer and ignition from the surroundings is called autoignition temperature.

Flash Point (FP): The lowest temperature at which a liquid gives off sufficient vapor to form an ignitable mixture with air near the surface of the liquid is called the flash point. At flash point, the combustion is brief.

Fire Point: The lowest temperature at which a liquid gives off sufficient vapor to form an ignitable mixture with air, and sustains combustion once ignited, is called the fire point.

Upper Flammability Limit (UFL): The highest concentration of a flammable fluid in air, above which the air-fuel mixture will not burn, is called the upper flammability limit. The fuel concentration, by definition, is expressed as follows.

$$UFL = \frac{\text{moles fuel}}{\text{moles fuel} + \text{moles air}}$$

Lower Flammability Limit (LFL): The lowest concentration of a flammable fluid in air below which the air-fuel mixture will not burn is called the lower flammable limit. The concentration of fuel is expressed as in UFL.

Lower Explosive Limit (LEL) and *Upper Explosive Limit (UEL)* are used interchangeably with LFL and UFL.

Explosion: The bursting or rupture of an enclosure or a container due to the development of internal pressure from a deflagration is called explosion.

Deflagration: Propagation of combustion zone front at a velocity that is less than the speed of sound in the unrecated medium is called deflagration.

Detonation: Propagation of combustion zone front at a velocity that is greater than the speed of sound in the unreacted medium is called detonation.

Combustible Dust: Any finely divided material, 420 microns or less in diameter (i.e., passing through a US # 40 sieve), that presents a fire or explosion hazard when dispersed and ignited in air or other gaseous oxidizer is called combustible dust.

Minimum Oxygen Concentration (MOC): The minimum percent of oxygen in air plus fuel required to propagate flame is called the MOC. This is also used to signify minimum *oxidant* concentration because substances other than oxygen can cause ignition. By definition, MOC is expressed by

$$MOC = \frac{\text{moles oxygen}}{\text{moles oxygen} + \text{moles fuel}}$$

It follows from the definition of MOC and LFL that

$$MOC = LFL \left(\frac{\text{stoichiometric moles of oxygen}}{1 \text{ mole of fuel}} \right)$$

The stoichiometric number of moles of oxygen needed to completely burn 1 mole of fuel to carbon dioxide and water can be determined from the stoichiometry of combustion reaction (parameter z in the following pages).

Minimum Ignition Energy (MIE): The minimum energy in millijoules required to start combustion is called MIE.

CLASSIFICATION OF LIQUIDS FOR THERMAL HAZARDS ANALYSIS

A *flammable liquid* is one having a closed cup flash point below 100°F (37.8°C) and a vapor pressure not exceeding 40 psia at 100°F. It is also known as a Class I liquid. Flammable liquids are subdivided into three classes.

	Class IA	Class IB	Class IC
Flash point °F:	<73	<73	≥73 < 100
Boiling point °F:	<100	≥100	<100
Venting device	CV & FA	CV or FA	CV or FA
for above-ground tanks:	Note 1	Note 1,2,3	Note 1,2,3

Notes:
(1) CV = Conservation vent, FA = Flame arrestor.
(2) Tanks of 3000 bbl capacity or less containing crude petroleum in crude producing areas, and outside aboveground atmospheric tanks under 23.8 bbl capacity shall be permitted to have an open vent.
(3) CV or FA may be omitted where conditions such as plugging, crystallization, polymerization freezing, etc., may cause obstruction resulting in the damage of the tank. Consideration should be given to heating the devices, liquid seal, or inerting.

A *combustible liquid* is one having a closed cup flash point at or above 100°F. Combustible liquids are subdivided into three classes.

	Class II	Class IIIA	Class IIIB
Flash point °F	>100 <140	≥140 <200	≥200

Aboveground tanks larger than 285 bbl capacity storing Class IIIB liquids and not within the diked area or the drainage path of Class I or Class II liquids do not require emergency relief venting for fire exposure per NFPA-30 code but may require consideration of other standards, such as API-520.

COMPUTATION OF FLASH POINT

To compute the flash point of a solution containing a single flammable or combustible liquid dissolved in inerts, do the following:

1. Find or experimentally determine the flash point of the flammable liquid in its pure state.

2. Find the vapor pressure of the flammable liquid at the flash point. Call it p_i.

3. Compute the pure component vapor pressure P^0_i by

$$P^0_i = p_i/x_i$$

where
   x_i is the concentration in mole fraction of the flammable liquid in solution

4. Determine the temperature at which the vapor pressure of the flammable liquid equals P^0_i. This temperature is the flash point of the solution.

CALCULATING FLAMMABILITY LIMITS

The flammability limits of a mixture of combustible gases may be reliably predicted by the Le Chatlier formulas.

$$LFL_{mixture} = \frac{1}{\frac{y_1}{LFL_1} + \frac{y_2}{LFL_2} + \cdots + \frac{y_n}{LFL_n}}$$

$$UFL_{mixture} = \frac{1}{\frac{y_1}{UFL_1} + \frac{y_2}{UFL_2} + \cdots + \frac{y_n}{UFL_n}}$$

where
   y = mole fraction of components in the mixture on a combustible basis or inert-free basis

This is calculated by dividing the % volume of the combustible component in the mixture by the % volume of all the combustible components. Subscripts denote the components. If the LFL_i and UFL_i in the denominator are entered as %, then the answer is also obtained as %. If the % volume of the total combustibles in the mixture, including the inerts, falls in the flammability range of the mixture (i.e., range of $LFL_{mixture}$ to $UFL_{mixture}$), then the mixture is flammable.

The dependence of flammability limit on temperature is expressed by the correlation of Zabetakis, Lambiris, and Scott:

Flammability Limit at T °C = Flammability Limit at 25 °C$[1 \pm 0.75 (T - 25)/\Delta H_c]$

where
- the + sign before 0.75 is to be used for UFL
- the − sign is for LFL
- ΔH_c is net heat of combustion in kcal/mol

Zabetakis also developed the correlation of flammability limit and pressure.

$$UFL_p = UFL + 20.6(\log P + 1)$$

where
- P = pressure in megapascals absolute = MPa, gauge + 0.101
- UFL = upper flammability limit at 1 atm
- Lower flammability limit has little sensitivity to pressure

STOICHIOMETRY OF COMBUSTION REACTION AND ESTIMATION OF FLAMMABILITY LIMITS

The stoichiometry of the combustion reaction of hydrocarbons containing C, H, and O may be represented by

$$C_m H_x O_y + zO_2 = mCO_2 + (x/2)H_2O$$

By stoichiometry

$$z = m + x/4 - y/2 = \text{moles of oxygen per mole of fuel}$$

For hydrocarbon vapors, Jones formulas may be used to estimate flammability limits.

$$LFL = 0.55\, C_{st}$$

$$UFL = 3.5\, C_{st}$$

where
- LFL and UFL are in % by volume
- C_{st} is the stoichiometric concentration of fuel in the combustion reaction, also in % by volume, and is defined by

$$C_{st} = \frac{(\text{moles fuel})100}{\text{moles air} + \text{moles fuel}}$$

$$= \frac{100}{\frac{\text{moles air}}{\text{moles fuel}} + 1}$$

$$= \frac{100}{\frac{\text{moles air}}{\text{moles oxygen}} \times \frac{\text{moles oxygen}}{\text{moles fuel}} + 1}$$

$$= \frac{100}{\frac{z}{0.21} + 1}$$

By combining the preceding equations, it can be shown that

$$LFL = \frac{55}{4.76m + 1.19x - 2.38y + 1}$$

$$UFL = \frac{350}{4.76m + 1.19x - 2.38y + 1}$$

INERTING AND PURGING

Inerting is the process of rendering a combustible mixture nonignitable by the addition of an inert gas. Most inerting methods use nitrogen, carbon dioxide, or steam as an inert gas.

Purging is the technique of reducing the initial concentration of a gas, most commonly oxygen.

Three common methods of purging are

1. Sweep through purging
2. Pressure purging
3. Vacuum purging

Sweep Through Purging

The purging gas is added to a vessel at one end and withdrawn at another end continuously. Assuming perfect mixing, with a continuous volumetric purge rate of Q (volume/time) through a container of volume V containing a toxicant, oxygen, or a combustible substance with a concentration C, which is to be reduced in time t, one can write a differential equation as follows:

$$-V\frac{dC}{dt} = QC$$

Separating variables and integrating, one gets the following:

$$t = \frac{V}{Q}\ln\left(\frac{C_0}{C}\right)$$

The preceding equation shows purging time for perfect mixing in the vessel. In a real situation, a mixing factor K is introduced for imperfect mixing. K ranges from 0.1 to 0.5 in most real situations.

$$t = \frac{V}{KQ}\ln\left(\frac{C_0}{C}\right)$$

where
 C_0 is the concentration in the beginning
 C = concentration at the end

Defining $N = Qt/V$ = number of changes of container volume, one can write

$$N = \frac{1}{K}\ln\left(\frac{C_0}{C}\right)$$

Pressure Purging and Vacuum Purging

Pressure and vacuum purging are batch operations. In pressure purging, a vessel is pressurized with an inert and then vented usually to atmospheric pressure. In vacuum purging, the contents of the vessel are withdrawn by a vacuum system until a predetermined vacuum is reached, followed by an injection of an inert to the vessel to bring the pressure to the initial condition. In both cases the operation may be repeated until the desired concentration is reached. Both methods may be evaluated by the same formulas.

$$n = \frac{\ln\left(\dfrac{y_0}{y_n}\right)}{\ln\left(\dfrac{P_h}{P_l}\right)} M_i = \frac{n(P_h - P_l)MV}{RT}$$

where

M = molecular mass of inert, lb/lbmol
M_i = mass of inert required for purging, lb
P_h = high pressure of the cycle, psia
P_l = low pressure of the cycle, psia
n = number of purging cycles required
R = gas constant, 10.731 (psia·ft^3/lbmol·°R)
T = temperature of operation, °R
V = volume of container, ft^3
y_0, y_n = concentration of oxygen or combustible substance in the beginning and after n purge cycles, in consistent units

LIMITING AND EXCESS REACTANT

In industrial practice exact stoichiometric amounts of materials are rarely used. An excess of stoichiometric amount is used to avoid an unsafe condition created by an unreacted energetic component or to consume a costly reactant completely. For example, excess air is used to avoid incomplete combustion of fuel. Two concepts should be understood in this area:

- Limiting reactant is the reactant that is present in a system in the smallest stoichiometric amount.

- Excess reactant is the reactant that is present in excess of the stoichiometric amount required by the limiting reactant.

$$\% \text{ Excess} = \frac{\text{moles in excess of what is required to react with limiting reactant}}{\text{moles required to react with the limiting reactant}} (100) \quad (100)$$

REFERENCES

1. Bodurtha, F.T., *Industrial Explosion Prevention and Protection*, New York, McGraw-Hill.
2. Crowl, D.A., and J.F. Louver, *Chemical Process Safety: Fundamentals with Applications*, Englewood Cliffs, NJ, Prentice Hall.
3. Das, D.K., and R.K. Prabhudesai, *Chemical Engineering License Review*, 2nd ed. Chicago, IL, Kaplan AEC Education, 2004.
4. *NFPA 69* 1992 edition

PROBLEMS

12.1 Estimate the flash point of a solution containing 50% by mole of ethanol in water. The flash point of pure ethanol is 13 °C. The vapor pressure-temperature relationship of ethanol is given by

$$\ln P_{mmHg} = 18.912 - \frac{3804}{T(K) - 41.68}$$

12.2 Estimate the MOC, LFL, and UFL of methanol in air.

12.3 Sixteen pounds per hour of methane is to be burned using air to form carbon dioxide and water. Calculate the amount of air necessary so that the supplied oxygen is 5% by volume in excess of what is needed for the stoichiometric amount. Assume composition of air as nitrogen (79% by volume) and oxygen (21%). The molecular weight of air is 28.97, and the molecular weight of methane is 16.

SOLUTIONS

12.1 The flash point is $13\,°C = 286$ K. The vapor pressure of ethanol at 286 K is

$$P_{ethanol\ @\ 286\ K} = e^{18.912 - \frac{3804}{286 - 41.68}} = 28.28 \text{ mmHg}$$

The pure component vapor pressure that corresponds to the partial pressure of 28.28 mmHg of a 50% solution of ethanol is given by

$$P^0_{ethanol} = 28.28/0.5 = 56.56 \text{ mmHg}$$

The temperature that corresponds to the vapor pressure of 56.56 mmHg of ethanol is

$$T = \frac{3804}{18.912 - \ln 56.56} + 41.68 = 297.38 K = 24.38\,°C$$

Therefore, the flash point of the solution is $24.38\,°C$.

12.2 Consider the stoichiometry of combustion reaction:

$$C_m H_x O_y + z O_2 = m CO_2 + (x/2) H_2 O$$

Since methanol CH_3OH may be considered $C_1H_4O_1$

$m = 1$, $x = 4$, $y = 1$, and hence $z = m + x/4 - y/2 = 1.5$. Therefore

$$C_{st} = \frac{100}{4.76 x 1 + 1.19 x 4 - 2.38 x 1 + 1} = 12.92\,\%$$

$$LFL = 0.55 C_{st} = 7.11\,\%$$

$$UFL = 3.5 C_{st} = 3.5 x 12.92 = 45.22\,\%$$

$$MOC = LFL \left(\frac{\text{stoichiometric moles of oxygen}}{1 \text{ mole of fuel}} \right) = (LFL) z$$
$$= 7.11 \times 1.5 = 10.67\,\%$$

12.3 Consider the stoichiometry of combustion of methane:

$$CH_4 + O_2 = CO_2 + H_2O$$

Basis: 16 lbs/hr of methane

lbmoles of methane/hr = 1

Limiting oxygen necessary = 1 lbmole/h

$$5 = \frac{\text{moles oxygen present} - \text{limiting oxygen moles}}{\text{limiting oxygen moles}} (100)$$

Moles oxygen present = 0.05 (limiting oxygen moles) + limiting oxygen moles = 0.05 (1) + 1 = 1.05

Air necessary to supply this oxygen = 1.05/0.21 = 5 lbmoles/h = 5(28.97) = 144.85 lbs/h

CHAPTER 13

Pollution Prevention

OUTLINE

FEDERAL POLLUTION PREVENTION ACT OF 1990 200

TERMINOLOGY 200
Acceptable Daily Intake (Dose) ■ Acid Rain ■ Activated Carbon ■ Activated Sludge ■ Biochemical Oxygen Demand (BOD) ■ Carcinogens ■ Chemical Oxygen Demand (COD) ■ Fugitive Emission ■ Greenhouse Effect ■ Hazardous Air Pollutant (HAP) ■ Lethal Concentration and Dose ■ Life Cycle Analysis ■ Mutagen ■ pH ■ Total Organic Carbon (TOC) ■ Volatile Organic Compounds (VOC)

OZONE, FRIEND AND FOE 203
Beneficial Ozone and Its Depletion ■ Harmful Ozone and Its Creation

ABATEMENT OF AIR POLLUTION 204
VOC Control Technologies ■ Non-VOC Control Technologies

ABATEMENT OF WATER POLLUTION 206

TREATMENT OF CONTAMINATED SOIL 207

REFERENCES 207

Pollution is a by-product of civilization because it is thermodynamically impossible to eliminate all wastes from the processes involved in the transformation of matter and energy required to sustain a civilized society. In the name of civilization and progress, however, humans have a tendency to rationalize defecation in their own nests.

Philosophically, pollution is misplacement of a substance in a concentration or amount that is directly or indirectly hazardous to health or life in any form. Technically, only when such misplacement violates the local or federal laws is it called pollution. However, misplacement of substances without violating laws does not necessarily mean a healthy atmosphere.

Pollution prevention should not be left as an end-of-pipe problem. It should be addressed at the stage of conceptual design. Careful consideration should be given to the waste treatment cost, which is affected by the choice of solvents, starting raw materials, operating parameters and their mode of control, and work practices, including detection, monitoring, reduction, and reporting. Most pollution can be prevented, or at least minimized, when the best available control technology is considered in the beginning of the conceptual design with support from management guided by enlightened self-interest.

FEDERAL POLLUTION PREVENTION ACT OF 1990

The central points of the Federal Pollution Prevention Act of 1990 can be summarized as follows:

1. Pollution should be prevented or reduced at the source wherever possible.
2. Pollution that cannot be prevented should be recycled in an environmentally safe manner whenever feasible.
3. Pollution that cannot be prevented or recycled should be treated in an environmentally safe manner whenever feasible.
4. Disposal or other release into the environment should be considered only as a last resort and should be conducted in an environmentally safe manner.

The Act defines the following categories of waste:

1. Municipal solid wastes such as paper, glass/metal, food, plastics, etc. Disposal involves landfill, incineration, recycling, etc.
2. Industrial hazardous wastes. Disposal involves recycle, landfill, incineration, land treatment, reuse as fuels, underground injection, waste piles, waste treatment and controlled discharge to natural water, and air.

TERMINOLOGY

Following are brief definitions of some key terminology of pollution prevention.

Acceptable Daily Intake (Dose)

It is the number of milligrams of a chemical per kilogram of body weight that, taken daily during an entire lifetime, appears to be without risk on the basis of all known facts at the time.

Acid Rain

All rainfalls are slightly acidic because of natural carbon dioxide. However, acid rain refers to the formation of acids due to the reaction of acidic oxides, such as oxides of sulfur and oxides of nitrogen with rainwater.

Activated Carbon

Carbon obtained from vegetable and animal sources and roasted in a furnace to develop high surface area ($1000 \ m^2/gm$) is called activated carbon.

Activated Sludge

Activated sludge is formed in the biological treatment of waste water that is mixed with biological culture (bacteria) in an aeration basin. Because of the growth of the bacteria and protozoa, part of the active sludge (about 90%) is incinerated or disposed of as landfill and balance recycled.

Biochemical Oxygen Demand (BOD)

It is the number of milligrams of dissolved oxygen consumed by one liter of waste for a specified incubation period in days at 20 °C. The incubation period is shown as a suffix. For example, BOD_5 denotes BOD with a 5-day incubation period.

Carcinogens

Any substance that induces cancer to man or animal is called a carcinogen. A material is considered a carcinogen if (1) it is certified as a carcinogen or potential carcinogen by IARC (International Agency for Research on Cancer), (2) it is listed as a carcinogen or potential carcinogen in the *Annual Report on Carcinogen*, (3) it is regulated by OSHA as a carcinogen, or (4) a positive study has been published. Some industrially known carcinogens are asbestos, aldrin, benzene, beryllium, cadmium, carbon tetrachloride, chloroform, *p*-dichlorobenzene, dieldrin, DDT, formaldehyde, hexamethylenediamine, selenium, tetraethyl lead, toluene-24-diamine, trichloroethylene, and vinyl chloride.

Chemical Oxygen Demand (COD)

It is the number of milligrams of oxygen that one liter of waste will absorb from a hot acidic solution of potassium dichromate.

The higher the values of BOD and COD, the more dangerous the waste is to aquatic life because the waste depletes oxygen from the water required for sustaining life. A substance with low BOD may mean that substance is not biodegradable.

Fugitive Emission

Fugitive emissions are unintentional releases from equipment such as pumps, valves, flanges, open-ended lines, filling losses, evaporation from sumps or ponds, and so on. The Environmental Protection Agency has five methods to estimate fugitive emission.

Greenhouse Effect

Combustion of fossil fuel (coal, oil, natural gas) generates carbon dioxide (about 20 billion tons/yr), which forms a layer in the earth's atmosphere. This layer traps heat radiating from the earth's surface thereby leading to the global warming effect. Potential effects may be melting of polar ice leading to flooding in coastal areas. Other contributors to the greenhouse effect are CFCs, methane, and ozone. The role of chemical engineers here is to develop alternative fuels, refrigerants, and propellants.

Hazardous Air Pollutant (HAP)

Pollutants are hazardous if they are carcinogens or if exposure to them may cause serious health problems. The Clean Air Act Amendment of 1990 includes 189 HAPs (Lipton and Lynch).[4]

Lethal Concentration and Dose

A concentration or dose that causes death is called the lethal concentration or dose. Lethal concentration, LC_{50}, is the concentration of the substance in mg per liter that causes death within 96 hrs to 50% of the test group of the most sensitive important species in the locality under consideration. Lethal dose, LD_{50}, is the dose of the substance in mg per kg of body weight of a specific animal (such as a rat) that causes death to 50% of the sample of the animals under test. A substance with LD_{50} greater than 7 g/kg is not considered harmful (Porteous).

Life Cycle Analysis

Life cycle analysis of a product considers all impacts on the environment starting from the procurement of the raw material to the generation, use, and disposal of the product.

Mutagen

Anything, such as a chemical or radiation, that changes the chromosomes of a cell so that the chromosomes of the daughter cells will be changed after cell division is called a mutagen. A mutagen may be carcinogen or teratogen (an agent that causes birth defects; dioxin, for example).

pH

The definition of pH is the negative logarithm to the base 10 of hydronium ion activity.

$$pH = -\log_{10}\left(a_{H_3O^+}\right) = -\log_{10}\left(\gamma C_{H_3O^+}\right) \cong -\log_{10} C_{H_3O^+} \cong -\log_{10} C_{H^+}$$

where

a = activity
γ = activity coefficient, usually 1 for dilute solution
C = concentration in mols per liter
H_3O^+ = hydronium ion
H^+ = hydrogen ion

The pH for water-based solutions is normally expressed in a scale of 0 (1 normal acid) to 14 (1 normal base). This scale is arbitrary, and therefore, pH could be negative (stronger than 1 normal acid) and also greater than 14 (stronger than 1 normal base). A pH of 7 indicates a neutral solution. Because of the base 10 logarithm scale, for example, a solution of pH at 6 is 10 times as acidic as a solution of pH at 7. The pH of pure water is 7 only at 25 °C, and it decreases as the temperature increases because of increased dissociation. The pH of a solution is affected by the nature of the solvent and even by neutral salts. This is because the activity coefficient is dependent on the ionic strengths of all ions, not just hydronium ions. Thus

$$\log_{10} \gamma_{H_3O^+} = \frac{-0.5 I^{0.5}}{1 + 3 I^{0.5}}$$

where

I = ionic strength = $1/2\sum(C_i Z_i^2)$, where the subscript denotes the ionic species, and Z denotes the ionic charge

Acceptable pH of various waters (according to Azad) are as follows:

Irrigation 4.5–9

Municipal secondary treatment waste and freshwater aquatic life and wildlife 6–9

Public supply and recreational water 5–9

Marine water 6.5–8.5

Total Organic Carbon (TOC)

It is number of milligrams of soluble organic carbon per liter of waste. Before the analysis, the inorganic carbon materials are removed by acidification and sparging, and insoluble solids are filtered. The test involves analysis of carbon dioxide generated by burning the sample.

Volatile Organic Compounds (VOC)

Any compound of carbon, *excluding* carbon monoxide, carbon dioxide, carbonic acids, metallic carbides or carbonates, and ammonium carbonate, that participates in atmospheric photochemical reactions is termed a VOC.

OZONE, FRIEND AND FOE

Ozone is a gas consisting of three atoms of oxygen. Depending on where it exists, it can be beneficial or harmful.

Beneficial Ozone and Its Depletion

Ozone that is created naturally at a distance of 15 to 20 miles above the surface of the earth protects life on the earth by absorbing a major portion of cancer-causing ultraviolet radiation and, in so doing, releases heat in the stratosphere. CFC (chlorofluorocarbons used as refrigerant, fire extinguisher, and propellant) escape undecomposed to the stratosphere, where, bombarded with ultra-violet radiation, they release halogens that attack ozone, breaking it down to oxygen. Nitrogen oxides released by supersonic jets also destroy ozone. Thus depletion of friendly ozone takes place, allowing more of the ultraviolet radiation to reach the surface of the earth.

The chemistry of stratospheric ozone involves a cyclical process of photochemical reaction of formation and decomposition of ozone:

(a) Formation of ozone

$$O_2(g) + h\nu \rightarrow 2\,O(g)$$
$$O(g) + O_2(g) + M(g) \rightarrow O_3(g) + M^*(g) + \text{heat}$$

(b) Decomposition of ozone

$$O_3(g) + h\nu \rightarrow O_2(g) + O(g)$$
$$O(g) + O(g) + M(g) \rightarrow O_2(g) + M^*(g) + \text{heat}$$

The chemistry of ozone depletion depends on the type of pollutant. With a CFC it is as follows:

(a) Photolysis or light-induced rupture of the carbon-chlorine bond and generation of atomic chlorine

$$CF_xCl_{4-x}(g) + h\nu \rightarrow CF_xCl_{3-x}(g) + Cl(g)$$

(b) Reaction of atomic chlorine with ozone leading to the depletion of the latter through regeneration of atomic chlorine

$$Cl(g) + O_3(g) \rightarrow ClO(g) + O_2(g)$$
$$ClO(g) + O(g) \rightarrow Cl(g) + O_2(g)$$

In the preceding equations, $h\nu$ represents photon energy (h = Planck's constant, ν = frequency of radiation), $M = N_2$, O_2, or H_2O, M^* = molecule M with excess energy.

Harmful Ozone and Its Creation

Although it is beneficial at the stratospheric level, ozone is dangerous at ground level. It is extremely reactive and toxic to breathe. It causes bronchitis and is extremely dangerous to asthma sufferers. Ozone is the key component of *smog*. Ozone is created by the oxides of nitrogen that are produced by automobiles, planes, chemical plants, burners, and so on. Oxides of nitrogen decompose by the bombardment of infiltrated ultraviolet radiation to nascent oxygen, which reacts with molecular oxygen to produce ozone. Ozone is a suicidal byproduct of civilization.

The chemistry of the formation of oxides of nitrogen (NO_x) and ozone in engines and furnaces are as follows:

$$N_2(g) + O_2(g) = NO(g) \quad \text{[Inside internal combustion engine or furnace]}$$
$$2NO(g) + O_2(g) = NO_2(g) \quad \text{[Inside IC engine, furnace and outside]}$$
$$NO_2(g) + h\nu \rightarrow NO(g) + O(g) \quad \text{[Outside air]}$$
$$O(g) + O_2(g) + M(g) \rightarrow O_3(g) + M^*(g)$$

ABATEMENT OF AIR POLLUTION

Let us look briefly at methods of abating both VOC and non-VOC air pollution.

VOC Control Technologies

- *Thermal oxidation (incineration).* By this technology, the VOC-laden air is burned at 1300–1800°F using supplementary fuel and air, if required, to produce water and carbon dioxide.

- *Catalytic oxidation (incineration).* This method is the same as thermal oxidation except a catalyst is used to lower the temperature of the reaction to 700–900°F, consequently reducing the fuel consumption.

- *Adsorption.* This technology uses a medium such as activated carbon or zeolite to adsorb VOC by weak intermolecular forces. The adsorbed VOC may be recovered by steam stripping or vacuum desorption and condensation.

- *Absorption.* By this method, a liquid solvent, such as water, high boiling hydrocarbons, caustic solutions (for acidic VOC), or acid, may be used in a venturi, packed/tray/spray column to capture VOC.

- *Condensation.* This technique uses low temperature coolants, such as brine, cryogenic fluid, or chilled water, in a surface condenser. Cryogenic fluids, like liquid nitrogen and carbon dioxide, can be directly injected into the VOC stream to effect condensation (direct contact condensation).

- *Waste heat recovery boiler.* In this process, the VOC-laden air is burned in boilers to generate steam.

- *Flare.* A flare, which is generally used to control pollution during process upsets, may also be used to burn VOC-laden air.

- *Biodegradation.* This process uses soil or compost beds containing cultured microorganisms to convert VOC into harmless components. The VOC-laden air must be dust free and humid.

- *Membrane separation.* Semipermeable membranes are used to separate VOC from air due to selective permeability of the gases.

- *Ultraviolet oxidation technology.* By this technology, the VOC is converted to carbon dioxide and water by oxygen-based oxidants, including ozone, peroxides, and radicals such as OH and O in the presence of ultraviolet light.

Non-VOC Control Technologies

Non-VOC air pollutants include carbon monoxide, carbon dioxide, halogens, halogen acids, nitrogen oxides, ammonia, and sulfur compounds. Many of the treatment technologies outlined for VOC also apply for pollution control of non-VOC.

Control of oxides of sulphur includes the following methods:

- Lime/limestone scrubbing
- Alkaline-fly-ash scrubbing
- Sodium carbonate/bicarbonate/hydroxide scrubbing
- Magnesium oxide scrubbing
- Sodium sulfite regenerative process (Wellman-Lord Process)
- Citrate process, which uses aqueous solution sodium sulfite, bisulfite, sulfate, thiosulfate, and polythionate buffered by citric acid
- Dimethyl aniline/xylidine process, which uses the compounds as a solvent to absorb sulfur dioxide
- Claus process, which uses hydrogen sulfide to react with sulfur dioxide to produce elemental sulfur

ABATEMENT OF WATER POLLUTION

Water pollutants are classifed as follows:

1. Floating pollutants, such as oil
2. Suspended pollutants, such as organic or inorganic solids
3. Dissolved pollutants, such as acids or organic and inorganic substances

Treatment processes for water pollution include the following:

1. Pretreatment, such as air floatation, pH adjustment, coagulation, precipitation, and flocculation.
2. Clarification, such as sedimentation and removal of suspended solids.
3. Filtration, such as cartridge filter or drum, and plate-and-frame filter.
4. High gradient magnetic separation (HGMS) for separation of magnetic, weakly magnetic, or nonmagnetic particles.
5. Aerated lagoons used for treating small quantities (<1 million gallons/day) of waste water with BOD (150–300 mg/L).
6. Biological treatment of organic wastewater using microorganisms that convert the organics into carbon dioxide, water, and methane gas. The most common types of biological processes are activated sludge (aerobic digestion, converts organic matter to carbon dioxide and water), trickling filter, fixed activated sludge treatment (a hybrid of activated sludge and trickling bed), anaerobic digestion (converts organic matter to carbon dioxide and methane), rotating biological contactor, and nitrification. Nitrification is a biological treatment involving ammonium waste, which is converted to nitrates by two types of bacteria: nitrosomonas, which oxidize ammonium to nitrite, and nitrobacter, which oxidizes nitrites to nitrate.

    Air in the aerobic digestion may be replaced by pure oxygen (UNOX process).

7. Carbon adsorption.
8. Ion exchange.
9. Reverse osmosis, electrodialysis, and ultrafiltration.
10. Incineration.
11. Stripping, as applied to ground water.
12. Liquid-liquid extraction.
13. Chlorination.
14. Ozone treatment for organic and inorganic waste.
15. Freeze concentration, evaporation, and crystallization.
16. Oxyphotolysis, a combination of ozone treatment and ultraviolet light used for toxic organics, such as malathion
17. Carbon-catalyzed hydrogen peroxide treatment to remove cyanide.

TREATMENT OF CONTAMINATED SOIL

There are two main methods of treating contaminated soil:

1. Bioremediation: treatment of contaminated soil with microorganisms
2. Incineration

REFERENCES

1. Azad, H. S., *Industrial Waste Water Management Handbook*, New York, McGraw-Hill Book Company.
2. Mukhopadhyay, N., and Moretti, E. C., *Current and Potential Future Industrial Practices for Reducing and Controlling Volatile Organic Compounds*, New York, American Institute of Chemical Engineers, .
3. Porteous, A., *Dictionary of Environmental Science and Technology*, New York, John Wiley & Sons, .
4. Lipton, S., and Lynch, J., *Handbook of Health Hazard Control in the Chemical Process Industry*, New York, John Wiley & Sons, Inc., pp. 106–112.

PROBLEMS

13.1 Calculate the biochemical oxygen demand in kg of 2000 m³ of waste with BOD_5 of 100.

13.2 Calculate the fugitive emission from a pump, inlet flange, and outlet flange when the screening concentration (ppmv) is 800.

Use the following correlations:

$$\text{Pump fugitive emission (kg/h)} = 1.333 \, (10^{-5.34})(\text{ppmv})^{0.898}$$
$$\text{Flange fugitive emission (kg/h)} = 0.918 \, (10^{-4.733})(\text{ppmv})^{0.818}$$

SOLUTIONS

13.1

$$\text{Biochemical oxygen demand} = 100\frac{\text{mg}}{\text{l}} \times 1000\frac{\text{l}}{\text{m}^3} \times 2\times 10^3\,\text{m}^3 \times \frac{1\,\text{kg}}{1\times 10^6\,\text{mg}}$$
$$= 200\text{ kg}$$

13.2

$$\text{Emission from the pump seals} = 1.22\,(10^{-5.34})(800)^{0.898} = 0.00246 \text{ kg/h}$$
$$\text{Emission from the pump flanges} = 2(0.918)(10^{-4.733})(800)^{0.818} = 0.00805 \text{ kg/h}$$
$$\text{Total emission} = 0.0105 \text{ kg/h}$$

CHAPTER 14

Distillation

OUTLINE

VAPOR-LIQUID EQUILIBRIA 212
Multicomponent and Binary Systems ■ Gibbs Phase Rule
■ Boiling Point and Equilibrium Diagrams (Binary Systems)
■ Enthalpy-Composition Diagrams (Binary Systems)

IDEAL SYSTEMS 215

RELATIVE VOLATILITY 216

NONIDEAL SYSTEMS 217
Van Laar Equations ■ Margules Equations

EQUILIBRIUM VAPORIZATION RATIOS
(EQUILIBRIUM CONSTANTS) K 218

BUBBLE POINT 218
Dew Point

FLASH (EQUILIBRIUM) DISTILLATION 219

DIFFERENTIAL BATCH DISTILLATION 220

FRACTIONAL DISTILLATION 220

DESIGN OF COLUMNS FOR SIMPLE BINARY SYSTEMS 222
Plate-to-Plate Calculations ■ Sorel's Method ■ Lewis
Method ■ Graphical Methods ■ Ponchon-Savrit Graphical
Method ■ McCabe-Thiele Method ■ Number of Stages
at Low or Very High Concentrations

REFERENCES 229

Distillation is the process of separating components from their liquid mixtures with the use of thermal energy as the separating medium and by selecting proper conditions of temperature and pressure. It takes advantage of the fact that at equilibrium, the concentration of the more volatile component in the vapor phase is greater than that in the liquid phase. The basic data required in solving distillation problems are the equilibrium composition relationships between the liquid and vapor phases of the system undergoing distillation. In this review, emphasis is placed on the binary (two-component) systems.

VAPOR-LIQUID EQUILIBRIA

Equilibrium is a state or condition of a system in which there is an absence of change as well as the absence of any tendency to change. Separation of components by distillation requires that the composition of vapor is different from the composition of the liquid when the two are in equilibrium. The basic data needed for distillation design calculations are the phase equilibria between the liquid and vapor phases. Such data may be obtained from the literature, by actual experimental determination, or by using theoretical methods.

Multicomponent and Binary Systems

A multicomponent system comprises more than two components. A binary system consists of only two components. Both multicomponent and binary systems can be either ideal or nonideal. Ideal behavior of systems is found at low pressures and normal distilling temperatures. Nonideal behavior is exhibited by systems consisting of substances of dissimilar nature or if the temperature and pressure conditions are severe.

In the following sections, we review vapor liquid equilibria in both binary and multicomponent systems as well as for ideal and nonideal systems.

Gibbs Phase Rule

A number of methods have been developed to present experimental equilibrium data in graphical form. These must be consistent with the number of variables involved. For equilibrium condition, the phase rule allows determination of independent variables. The phase rule at constant pressure and temperature for physical (nonreacting) systems is as follows:

$$F = C + 2 - P$$

where
  F = degrees of freedom
  C = number of independent components
  P = number of distinct phases in the system

For a binary system at equilibrium comprising two phases (vapor and liquid), $C = 2$ and $P = 2$. Therefore, $F = 2$. Thus fixing two variables determines binary system equilibrium.

A ternary system with two phases in equilibrium has $F = 3$ and will be defined by three variables. For an n component system, and two phases in equilibrium, there are n degrees of freedom.

Boiling Point and Equilibrium Diagrams (Binary Systems)

For binary systems, the differences in compositions of the liquid and vapor phases are graphically represented by either boiling point (temperature-composition) or equilibrium (composition only) diagrams. In the former, boiling point temperature is plotted against liquid and vapor compositions, whereas in the latter, the equilibrium vapor phase compositions of more volatile component are plotted against the corresponding liquid phase compositions as in Figure 14.1a and b.

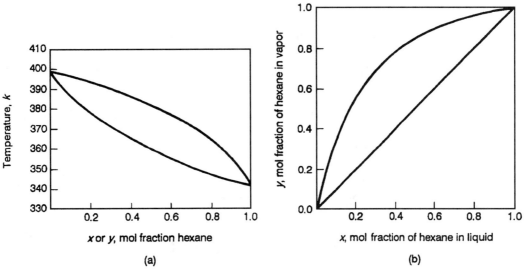

Figure 14.1 (a) T-x-y diagram; (b) equilibrium diagram

liquid and vapor mixture at equilibrium, the ratio of the moles of the liquid phase to moles of the vapor phase is given by the inverse lever rule as given in the following:

$$\frac{L}{V} = \frac{y_{AV} - x_{AF}}{x_{AF} - x_{AL}}$$

where
- L = mols of liquid phase
- V = mols of vapor phase in equilibrium with the liquid
- y_{AV} = mol fraction of component A in vapor phase
- x_{AF} = mol fraction of component A in the liquid feed before vaporization
- x_{AL} = mol fraction of component A in the liquid phase at equilibrium

For binary systems that obey Raoult's law and the vapor obeying Dalton's law of partial pressures, it is possible to calculate the boiling point and equilibrium diagrams from the vapor pressures of the pure components.

Enthalpy-Composition Diagrams (Binary Systems)

The enthalpy-composition diagram is a plot of vapor and liquid enthalpies versus the vapor and liquid compositions. This diagram must be constructed for a temperature range covering the two-phase region at the pressure of distillation. Since enthalpy has no absolute value, a reference state is needed to compute enthalpy changes. This is usually taken as 77°F or 25°C because most heats of mixing data are available at this temperature. In case one of the components is water, it is convenient to use 32°F or 25°C to calculate relative enthalpies. Once the basis is chosen, one can compute the relative enthalpies of liquid and vapor mixtures at a given pressure in the following manner.

$$\text{Enthalpy of liquid mixture, } \hat{h}_{\text{mixture}} = x_A \hat{h}_A + x_B \hat{h}_B + \Delta \hat{H}_{\text{mixing}}$$

where

$\hat{h}_{mixture}$ = relative enthalpy of liquid mixture per unit mass

$\hat{h}_A, \hat{h}_B$ = enthalpies of pure components (liquid state) per unit mass relative to reference temperature

$\Delta \hat{H}_{mixing}$ = heat of mixing (solution) per unit mass at the given temperature

The heat of mixing is usually available at 77 °F (25 °C), therefore if the reference temperature is different from this, both $\hat{h}_A$ and $\hat{h}_B$ need to be calculated at 77 °F (25 °C) by the following equation

$$(\hat{h}_A)_t - (\hat{h}_A)_{ref} = \int_{t_{ref}}^{t=77°F} C_{PA} dt$$

which reduces to $(\hat{h}_A)_t = \int_{t_{ref}}^{t=77°F} C_{PA} dt \quad \text{if} \quad (\hat{h}_A)_{ref} = 0$

In a similar manner, $(\hat{h}_B)_t = \int_{t_{ref}}^{t=77°F} C_{PB} dt \quad \text{if} \quad (\hat{h}_B)_{ref} = 0$

Once $\hat{h}_A$ and $\hat{h}_B$ at 77 °F (25 °C) are calculated, they are used to calculate $\hat{h}_{mixture}$ at any composition of liquid at 77 °F (25 °C). When this is done, relative enthalpies of liquid mixtures at any other temperature can be calculated by

$$\hat{h}_{\text{mixture at any } t} = \hat{h}_{\text{mixture at } 77°F=25°C} + \int_{77°F=25°C}^{T} C_{Px} dt$$

where

C_{Px} is average heat capacity of the liquid mixture at concentration x

After computing relative enthalpies of liquid mixtures at the bubble point temperatures, the next step is to calculate relative enthalpies of vapors in equilibrium with saturated liquid mixtures at their respective bubble point temperatures. The following data are required to make these calculations

- Latent heats of vaporization of the two pure components
- Heat of mixing data for vapors
- Heat capacities of pure liquids and vapors
- Bubble point temperatures as a function of composition
- Dew point temperatures of the vapors as a function of composition
- Vapor-liquid equilibrium data

The heats of mixing of vapors are usually negligible and ignored. Alternatively, latent heats of vaporization of a given composition can be used to construct the enthalpy curve of saturated vapor. For practical purposes the latent heat of a liquid mixture can be calculated by the equation

$$\lambda_{mixture} = x_A \lambda_A + x_B \lambda_B$$

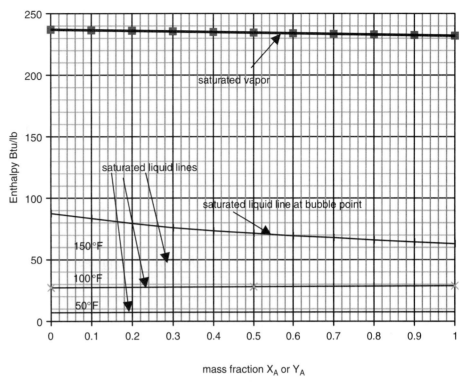

Figure 14.2 Enthalpy vs. composition diagram

Then enthalpy of the saturated vapor of the same composition as the liquid is given by

$$\hat{H}_V = \hat{h}_{\text{mixture at bubble point temperature}} + \lambda_{\text{mixture}}$$

Enthalpy of a saturated vapor in equilibrium with the saturated liquid can be calculated using the composition of the vapor in equilibrium with the liquid. Thus if y_A and y_B are equilibrium compositions of components A and B, then the enthalpy of equilibrium vapor will be given by the equation

$$\hat{H}_V = y_A[C_{PA}(t_V - t_{\text{ref}}) + \lambda_A] + y_B[C_{PB}(t_V - t_{\text{ref}}) + \lambda_B]$$

In making enthalpy calculations, it should be remembered that the *enthalpy is a state property* and any convenient path may be chosen to arrive to that state.

When no heat of solution data are available, and only pure component data are used to construct the enthalpy-composition diagram, the saturated vapor and saturated liquid enthalpy lines will be straight. An example of an enthalpy composition diagram is shown in Figure 14.2.

IDEAL SYSTEMS

Ideal liquid mixtures obey *Raoult's law*, which relates the partial pressure of a component in the vapor phase to the liquid phase composition by the expression

$$p_i = x_i P_i$$

Combining Dalton's law referred to in Chapter 2 with Raoult's law, the following expression can be obtained to describe mixtures of ideal vapors and liquids in equilibrium.

For a single component i, $y_i P_t = x_i P_i = p_i$

For a multicomponent system,

$$P_t = \sum_1^n P_i = \sum_1^n y_i P_t = \sum_1^n x_i P_i$$

$$y_i = \frac{x_i P_i}{\sum_1^n x_i P_i}$$

where
- p_i = partial pressure of component i in the vapor phase
- x_i = mole fraction of component i in liquid phase
- P_i = vapor pressure of pure component i at the system temperature
- y_i = mole fraction of component i in vapor phase
- P_t = total pressure of the system

Henry's law applies to the vapor pressure of a solute in a dilute solution. Raoult's law applies to the vapor pressure of the solvent. It states that the partial pressure of the solute is proportional to its concentration in the solution. This law may be expressed by the expression

$$p_a = k x_a$$

where
- p_a = partial pressure of the solute
- x_a = mol fraction of the solute in solution
- k = a constant that is to be determined experimentally

RELATIVE VOLATILITY

The relative volatility α_{ij} of a component i with respect to component j in a mixture is defined by the relation

$$\alpha_{ij} = \frac{y_i / x_i}{y_j / x_j}$$

For an ideal mixture, it is equal to the ratio of the vapor pressures or $a_{ij} = P_i / P_j$. For a *binary (two-component) system*, $y_i = 1 - y_j$ and $x_i = 1 - x_j$. Therefore

$$\alpha_{ij} = \left(\frac{y_i}{1 - y_i} \right) \left(\frac{1 - x_i}{x_i} \right)$$

In general

$$a = \left(\frac{y}{1 - y} \right) \left(\frac{1 - x}{x} \right)$$

from which

$$y = \frac{ax}{1+(a-1)x} \quad \text{and} \quad x = \frac{y}{a+y(1-a)}$$

In terms of vapor pressures of two components in a binary system, the preceding relations can be expressed as

$$y_1 = \frac{P_1 x_1}{P_1 x_1 + P_2(1-x_1)} = \frac{P_1 x_1}{P_t} \quad \text{and} \quad x_1 = \frac{P_t - P_2}{P_1 - P_2}$$

where

P_1 and P_2 are the vapor pressures of the components 1 and 2, respectively

NONIDEAL SYSTEMS

If the components show strong interactions (physical or chemical) then the boiling point and equilibrium diagrams for binary systems are quite different from those of the ideal systems (Van Winkle, Das and Prabhudesai, Henley, Perry, Treybal, Smith and Van Ness). Azeotropic mixtures have a critical liquid composition at which the liquid and vapor compositions are the same. In all systems (both multicomponent and binary), the partial pressures of components of a real solution will deviate from those calculated by Raoult's law. Nonideal behavior may exist in either or both phases. For system pressures close to atmospheric, the real behavior may be accounted for by including a correction factor γ_i, called the activity coefficient in Raoult's law. This results in the following relation

$$p_i = \gamma_i x_i P_i$$

The product $\gamma_i x_i$ is termed the activity a_i of the component i. The activity coefficient γ_i varies with temperature and composition and can be >1 or <1. By incorporating activity and fugacity coefficients γ_i and $\hat{\phi}_i$ in Raoult's law, the following relation results:

$$\hat{\phi}_i y_i P_t = \gamma_i x_i P_i$$

With the use of this relation, the relative volatility in a nonideal system is shown to be

$$a_{ij} = \frac{y_i/x_i}{y_j/x_j} = \frac{\gamma_i P_i \hat{\phi}_j}{\gamma_j P_j \hat{\phi}_i}$$

Many empirical and semiempirical equations have been proposed to predict the variation of the activity coefficients with temperature and composition. For nonideal binary systems, van Laar and Margules equations are widely used. They are described briefly next.

Van Laar Equations

These are as follows:

$$T \ln \gamma_1 = \frac{B}{[1 + A(x_1/x_2)]^2} \quad \text{and} \quad T \ln \gamma_2 = \frac{AB}{(A + x_2/x_1)^2}$$

where

A and B are constants that are assumed to be independent of temperature

At the azeotropic composition

$$x_i = y_i \quad \text{and} \quad \gamma_i = P_t/P_i, \text{ Also } \gamma_i = y_i P_t/x_i P_i$$

Therefore, knowing the azeotropic composition, the constants A and B can be calculated.

If the temperature variable is not included, another more frequently used van Laar equations model results. These equations are

$$\ln \gamma_1 = A_{12}\left\{1 + \frac{A_{12}x_1}{A_{21}x_2}\right\}^{-2} \quad \text{and} \quad \ln \gamma_2 = A_{21}\left\{1 + \frac{A_{21}x_2}{A_{12}x_1}\right\}^{-2}$$

where A_{12} and A_{21} are van Laar constants.

Margules Equations

These are given by

$$\ln \gamma_1 = x_2^2[A + 2x_1(B-A)] \quad \text{and} \quad \ln \gamma_2 = x_1^2[B + 2x_2(A-B)]$$

For a detailed treatment of activity coefficients and their estimation, the student is referred to other thermodynamic and distillation texts (Van Winkle, Henley, Perry, Treybal, Smith and Van Ness, Hougen, Robinson and Gilliland).

EQUILIBRIUM VAPORIZATION RATIOS (EQUILIBRIUM CONSTANTS) K

The K value is extremely useful in hydrocarbon distillation, absorption, and stripping calculations. It is defined by the following relation:

$$K_i = \frac{y_i}{x_i}$$

In terms of vapor pressure, activity coefficient, and the fugacity coefficient, K_i is given by

$$K_i = \frac{\gamma_i P_i}{\hat{\phi}_i P_t}$$

For an ideal system, both γ_i and $\hat{\phi}_i$ are unity and K_i reduces to $K_i = P_i/P_t$

BUBBLE POINT

The bubble point of a liquid mixture is the temperature at which it begins to boil at a given pressure such that the vapor is in equilibrium with the liquid. For a system containing n components, the equilibrium condition can be expressed by the following relation:

$$\sum_1^n K_i x_i = \sum_1^n y_i = 1.0$$

For ideal systems

$$\sum_1^n y_i = \sum_1^n \frac{x_i P_i}{P_t} = 1.0$$

Dew Point

Dew point is the temperature at which the vapor when cooled begins to condense at a given pressure. This can be expressed in terms of equilibrium relationships as follows:

$$\sum_1^n \frac{y_i}{K_i} = \sum_1^n x_i = 1.0 \text{ for ideal systems}$$

$$\sum_1^n x_i = \sum_1^n \frac{y_i}{P_i} P_t = 1.0 \text{ vapor-liquid}$$

Equilibrium data for *ideal multicomponent* systems can be calculated in the same way as those for binary systems. Four types of equilibrium data calculations are encountered when determining the bubble and dew point temperatures. These are as follows:

1. Liquid composition and temperature are known.
2. Vapor composition and temperature are known.
3. Liquid composition and total pressure are known.
4. Vapor composition and total pressure are known.

Except in case 1, the calculations require trial-and-error solution to fix the unknown variable (either pressure or temperature).

FLASH (EQUILIBRIUM) DISTILLATION

This involves vaporization of a fraction of a batch of liquid mixture, keeping the liquid and the vapor in contact so that the vapor is in equilibrium with the liquid. The vapor is then withdrawn and condensed. This is a batch operation. In continuous flash distillation, the vapor and liquid are withdrawn from the flash chamber continuously. For a binary system, the process calculations can be done as follows:

Overall material balance: $F = V + L$
Component material balance: $Fx_F = Vy_i + Lx_i$

These two equations in conjunction with the equilibrium diagram allow us to determine the relative amounts of vapor and liquid produced. The energy balance for the process can be written as follows:

$$Fh_F = VH_v + Lh_L$$

where

F = moles of feed
V = moles of flashed vapor
L = moles of undistilled liquid
h_F, H_v, and h_L = molar enthalpies of feed, vapor, and the condensed liquid

For a *multicomponent* mixture, the following equations apply:

$$x_i = \frac{x_{Fi}/L}{K_i V/L + 1} \quad \text{or} \quad x_i = \frac{x_{Fi}/V}{K_i + L/V}$$

Since $y_i = K_i x_i$,

$$y_i = \frac{K_i x_{Fi}/V}{K_i + L/V}$$

and either $\sum x_i = 1.0$ or $\sum y_i = 1.0$. If one mole of the feed liquid is considered and f is the fraction vaporized, then $(1 - f)$ is the liquid left behind. Then by material balance

$$xF_i = fy_i + (1 - f)x_i$$

from which

$$y_i = -\frac{1-f}{f}x_i + \frac{x_{Fi}}{f}$$

For multicomponent (>2 components) mixtures, a trial-and-error solution is required. In the case of binary systems, the fraction vaporized is given by

$$f = \frac{x_{Fi}(K_1 - K_2)/(1 - K_2) - 1}{K_1 - 1}$$

DIFFERENTIAL BATCH DISTILLATION

In this type of distillation, the liquid mixture to be distilled is charged to a still and heated. The vapor is withdrawn and condensed continuously until a desired quantity of the distillate is obtained. The boiling point of the liquid in the still gradually increases as the lower boiling components vaporize. The Rayleigh equation relating final moles in the still W_2 to the initial charge W_1 moles of feed liquid is

$$\ln\frac{W_1}{W_2} = \int_{x_{i2}}^{x_{i1}} \frac{dx_i}{y_i - x_i}$$

For a binary system, the above equation can be integrated graphically when x-y data are available. If relative volatilities α are constant, the direct solution of the equation is

$$\ln\frac{W_1}{W_2} = \frac{1}{\alpha - 1}\left(\ln\frac{x_{i1}}{x_{i2}} + \alpha\ln\frac{1 - x_{i2}}{1 - x_{i1}}\right)$$

For nonideal systems,

$$y_i = \frac{\gamma_i x_i P_i}{\hat{\phi}_i P_t} = K_i x_i \text{ and}$$

$$\ln\frac{W_1}{W_2} = \int_{x_{i2}}^{x_{i1}} \frac{dx_i}{x_i(K_i - 1)} = \int_{x_{i2}}^{x_{i1}} \frac{dx_i}{x_i(\gamma_i P_i/\hat{\phi}_i P_t - 1)}$$

FRACTIONAL DISTILLATION

When higher product purities and large-scale operation are required, continuous fractional distillation is employed. This involves contacting vapor and liquid in more than one stage, such as in a fractionating column (Figure 14.3a). Industrial

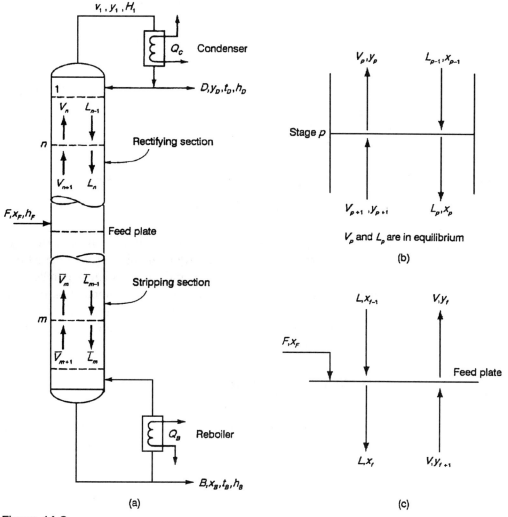

Figure 14.3

equipment is of two types: (1) plate columns and (2) packed columns. Examples of plate columns are bubble cap columns and sieve plate columns.

The basic model used in the analysis of a fractionator is the *equilibrium stage* (Figure 14.3b). This means at steady state operation of the column, the liquid and vapor leaving the stage are in equilibrium. The number of *theoretical stages* required for a given separation is first computed. The actual number of stages is then determined by using a plate efficiency factor.

A portion of the condensed distillate called *reflux* is returned to the top plate in the column. This causes vaporization of the more volatile (light) component at each stage and condensation of the less volatile (heavy) component in the top or above the feed portion of the column. This enrichment of the more volatile component is called *rectification* and the portion of the column where it occurs is termed the *rectification section*.

In the section below the feed plate, hot vapor from the reboiler passes through each stage causing vaporization of the more volatile or light component and condensation of the less volatile or heavy component. Thus the more volatile component is stripped from the liquid. This process is called *stripping* and the portion of the column where it is done is termed the *stripping section* of the column.

In multicomponent distillation, the light (more volatile) and the heavy (less volatile) components between which the required separation is specified are termed the *key components* (*light* key and *heavy* key). For the design of a column, the number of theoretical stages required for a given separation is first computed. The actual number of stages is then determined by using a *plate efficiency factor*. For the FE exam, only short, simple questions are asked and they generally refer to binary systems especially the McCabe-Thiele graphical method. We will therefore restrict our detailed review of design calculations for a distillation column to the binary systems.

DESIGN OF COLUMNS FOR SIMPLE BINARY SYSTEMS

Calculation of equilibrium stages is an important factor in the distillation column design because the number of stages required influences the size of the column and its cost. Equilibrium stage calculations for a multicomponent system are much more complicated and will not be covered in this review. Calculation of equilibrium stages for a binary system is comparatively simpler. A number of methods, analytical and graphical, have been developed for computation of equilibrium stages in columns for binary systems. The most important of these are as follows: analytical methods by Sorel and Lewis, graphical methods, such as by McCabe-Thiele, and the enthalpy-composition diagram.

Plate-to-Plate Calculations

One method of determining the number of equilibrium stages is to make plate-to-plate calculations. Conditions at the two ends (top and bottom) of the column are known. Hence either the top or bottom can be selected as the starting point for the calculations.

Sorel's Method

This method does plate-to-plate calculations without the assumption of constant molal overflow. It uses material and enthalpy balances together with equilibrium calculations to determine the flows of liquid and vapor from the plate, temperature of the plate, and the composition of each stream from the plate. The method assumes the operating pressure, the reflux ratio, temperature or enthalpy of the reflux stream, and the use of a total condenser. These calculations are rigorous, tedious, and time consuming for hand calculations. If relevant data are available, this method is suitable for high-speed computer calculation.

Lewis Method

This method also uses plate-to-plate calculations but simplifies them by assuming constant molal overflow from stage to stage in both the rectifying and stripping sections. This is equivalent to assuming equimolal latent heats and heat capacities and no heat of mixing. He also assumed the reflux L_0 to be a saturated liquid. Even then, like Sorel's method, it is tedious and time consuming.

Graphical Methods

Graphical methods are developed to determine the equilibrium stages because of the tedious trial and error calculations for each plate involved in plate-to-plate calculations. Two important methods are Ponchon-Savrit and McCabe-Thiele (Van Winkle).

Ponchon-Savrit Graphical Method

The Ponchon-Savrit graphical method uses an enthalpy-composition diagram for solving material and energy balances for binary systems. For the construction of ideal stages on this diagram, no simplifications, such as assumption of constant molal overflow, are made. The Ponchon-Savrit method will not be reviewed further here.

McCabe-Thiele Method

This method is based on the Lewis modification of the Sorel method. To simplify calculations, it makes a number of assumptions, which are the following:

1. Constant molal overflow in the rectifying section as well as in the stripping section, which means equimolal latent heats and heat capacities and no heat of mixing.
2. The reflux L_0 to the top plate in the column is a saturated liquid.
3. The column pressure is constant.
4. The reflux ratio is also assumed.

Material and energy balances for a binary system are shown in Figure 14.3a

$$\text{Overall material balance: } F = D + B$$
$$\text{Component } A \text{ balance: } Fx_F = Dx_D + Bx_B$$

where
F = feed to the column, moles/h
D = distillate, mole/h
B = bottoms product
X_F = mole fraction of component A in feed
X_D = mole fraction of component A in distillate
X_B = mole fraction of component A in bottoms product

Rectifying Section

With reflux at its bubble point (total condenser), the operating line equation for the rectifying section of the column can be derived and is as follows

$$y_{n+1} = \frac{L}{V}x_n + \frac{D}{V}x_D$$

where
L = liquid molar overflow from stage to stage, mol/h
V = vapor molar flow from stage to stage, mol/h
D = distillate product, moles/h
x_D = mole fraction of more volatile component in the distillate
x_n = mole fraction of component more volatile component in the liquid leaving stage n
y_{n+1} = mol fraction of more volatile component in the vapor leaving stage n

The equation can also be written in terms of external reflux ratio $R = L_0/D = L/D$ as

$$y_{n+1} = \frac{R}{R+1}x_n + \frac{x_D}{R+1}$$

The slope of the operating line is

$$\frac{L}{V} = \frac{R}{R+1}$$

and the intercept of the line is $x_D/(R+1)$.

Stripping Section

The corresponding equation of the operating line for stripping or bottom section of the column below the feed stage is

$$y_{m+1} = \frac{\overline{L}}{\overline{V}}x_n - \frac{B}{\overline{V}}x_B$$

where
$\overline{L}$ = liquid overflow in the stripping section, moles/h
$\overline{V}$ = vapor flow from stage to stage in the stripping section, moles/h

Since $\overline{V} = \overline{L} - B$, the preceding equation becomes

$$y_{m+1} = \frac{\overline{L}}{\overline{L}-B}x_n - \frac{B}{\overline{L}-B}$$

The intersection of the operating lines occurs at a point, which is determined by the feed composition and material and energy balance at the feed plate. Using material balance at the feed plate (Figure 14.3c)

$$F + L + \overline{V} = V + \overline{L}$$

Now

$$\overline{L} = L + qF \quad \text{or} \quad V = \overline{V} + (1-q)F$$

where q is defined as:

$$q = \frac{\overline{L}-L}{F} = \frac{\text{heat to convert 1 mole of feed to saturated vapor}}{\text{latent heat of vaporization per mole of feed}}$$

The equation of the stripping section operating line in terms of q is

$$y_{m+1} = \frac{L+qF}{L+qF-B}x_m - \frac{B}{L+qF-B}x_B$$

The equation of the q line can be obtained as

$$y = \frac{q}{q-1}x - \frac{x_F}{q-1}$$

The q line has a slope of $q/(q-1)$ and terminates at x_F on the 45° line in the equilibrium diagram. It also locates the intersection point of the operating lines for all values of q.

q Values and q-Line Position

The position of the q line on the equilibrium diagram depends upon its value, which is determined by thermal condition of the feed. The calculation of q is shown in Table 14.1. The effect of the q value on the slope of the q line is shown in Figure 14.4a and the method of the McCabe-Thiele graphical solution is shown in Figure 14.4b.

Table 14.1 Calculation of q and Slope of q line

Condition of Feed	Calculation of q	q	Slope of q Line $q/(q-1)$
Subcooled liquid	$q = 1 + C_{PL}(t_B - t_F)/\lambda$	$q > 1.0$	+
Feed at bubble point	$q = 1.0$	$q = 1.0$	∞
Partially vaporized feed	$q = f_L \lambda / \lambda = f_L$	$0 < q < 1.0$	−
Feed at dew point (saturated vapor)	$q = 0$	$q = 0$	0
Feed superheated vapor	$q = \dfrac{C_{PV}(t_D - t_F)}{\lambda}$	$q < 0$	+

In Table 14.1, the variables are as follows:

C_{PL} = molar specific heat of liquid
C_{PV} = molar specific heat of superheated vapor feed
t_D = dew point temperature
t_B = boiling point
t_F = feed temperature
f_L = molar liquid fraction
λ = latent heat of vaporization per mol

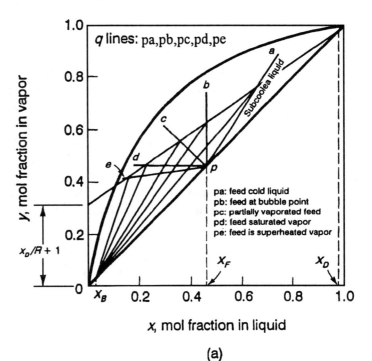

(a)

Figure 14.4

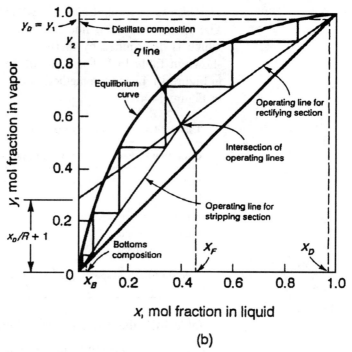

Figure 14.4 (continued)

Feed Plate Location

The optimum location of the feed plate gives the least number of equilibrium stages required for a specified separation and reflux ratio. Least number of stages is obtained if the operating line switch from rectifying to stripping section is made at the intersection of the two operating lines.

Minimum Reflux

Minimum reflux ratio corresponds to an infinite number of equilibrium stages. It is a conceptual limit never actually attainable. As the reflux ratio is decreased, the intersection point of the operating lines gets closer to the equilibrium line until finally the intersection point of the operating line and the q line lies on the equilibrium curve (Figure 14.5a) or one of the operating lines becomes tangent to the equilibrium curve (Figure 14.5b). In both cases, a pinch point is reached and an infinite number of equilibrium stages is required. These conditions correspond to minimum reflux, which can be calculated from the slope of the critical operating line or from the y-axis intercept as follows:

$$m = \frac{x_D - y_P}{x_D - x_P} = \left(\frac{L}{V}\right)_{min} = \frac{R_{min}}{R_{min} + 1} \quad \text{and} \quad R_{min} = \frac{m}{1-m}$$

Total Reflux

At total reflux, the operating lines for both sections of the column coincide with the 45° line of the equilibrium diagram. Then the number of equilibrium stages required is minimum, but there is no product. The minimum number of theoretical stages can be obtained graphically on the x-y diagram between compositions x_D and x_B using the 45° line as the operating line in both the rectifying and stripping

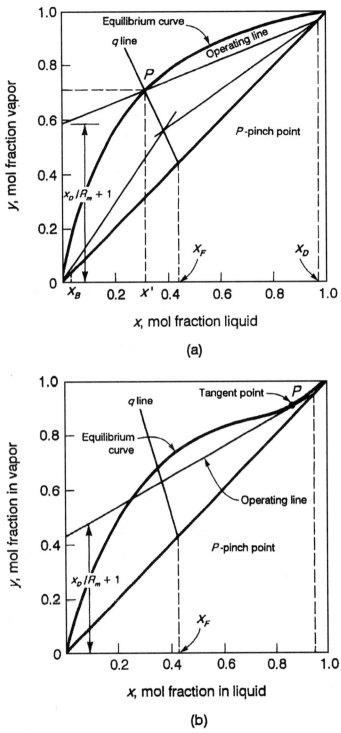

Figure 14.5

sections. For ideal mixtures and constant relative volatility, the minimum number of theoretical plates N_{min} can be calculated by the *Fenske equation*.

$$N_{min} = \frac{\log[x_D(1-x_B)/x_B(1-x_D)]}{\log \alpha_{AB}} - 1$$

If α changes moderately over the column, an average relative volatility can be used as follows:

$$\alpha_{av} = \sqrt{\alpha_{top}\alpha_{bottom}}$$

Partial Condenser

A partial condenser is one in which a portion of the vapor is condensed. These are used when a vapor product is desired or if the temperature required to condense the vapor is too low. Equilibrium condensation occurs if the time of contact of the liquid and vapor is adequate. In this case, the condenser is equivalent to a theoretical equilibrium stage. In the design of new equipment, it is safer to ignore the enrichment by the partial condensation and include an additional theoretical tray in the column itself. (Note: Usually the exam will assume a total condenser.)

Reboiler

The reboiler is usually considered as one equilibrium stage. However, it depends upon the configuration of the reboiler. If the liquid is partially vaporized in the reboiler and fed to the bottom of the column, a vapor and liquid in equilibrium are produced. In this case, the reboiler can be considered equivalent to one equilibrium stage. When a reboiler is used to boil a saturated liquid to a saturated vapor, no separation occurs and the reboiler is equivalent to a zero equilibrium stage. In the case of a thermo-siphon reboiler, the effect is of a fraction of an equilibrium stage. For safety in design, the thermo-siphon reboiler is counted as zero stage.

Optimum Reflux Ratio

As the reflux ratio is increased from the theoretical minimum, the total cost of running a distillation system consisting of the depreciation of the installed equipment and the operating cost of coolant, heating medium, and electricity first decreases, reaches a minimum value, and then increases. The reflux ratio at which the total cost is minimum is called the optimum reflux ratio. In actual practice, a reflux ratio equal to 1.2 to 2 times the minimum is used. The total cost is not very sensitive in this range and a better flexibility in operation is obtained.

Column Efficiency

To obtain actual number of trays required in a column, a tray efficiency factor is required. Murphree plate efficiency is calculated by

$$E_0 = \frac{y_n - y_{n+1}}{y_n^* - y_{n+1}}$$

where

E_0 = plate efficiency factor
y_n^* = vapor composition in equilibrium with L_n
y_n, y_{n+1} = actual vapor compositions from nth and $(n+1)^{th}$ plates

Plate efficiencies are to be determined experimentally but some prediction methods are also available. O'Connell gave an empirical correlation of plate efficiency as a function of feed viscosity and the relative volatilities of the key components. (Exam tip: Divide the theoretical number of plates by the efficiency to get the actual number of plates.) Bakowski's semitheoretical method and AIChE predictive method both give overall column efficiency. These are already referred to in Chapter 10.

Column Energy Balance

Neglecting the heat loss, the overall energy balance around the whole column gives

$$Fh_F + Q_B = Dh_D + Bh_B + Q_C$$

Condenser duty is given by $Q_C = D(R + 1)(H_1 - h_D)$

where

  h and H are enthalpies per unit mol
  Subscripts B, D, F, and 1 indicate bottom, distillate, feed, and tray 1, respectively

If condensate is at its bubble point (no subcooling)

$$Q_C = D(R + 1)\lambda$$

where

  λ is the latent heat of condensation of overhead vapor per mol

If t_F is taken as the datum temperature, the feed is 100% liquid, and t_D and t_B are the distillate and bottoms product temperatures, respectively.

$$h_F = 0, \quad h_D = C_{PD}(t_D - t_F), \quad \text{and} \quad h_B = C_{PB}(t_B - t_F)$$

where

  C_{PD} and C_{PB} are the molar specific heats of the distillate and bottoms product

Then reboiler duty is given by

$$Q_B = DC_{PD}(t_D - t_F) + BC_{PB}(t_B - t_F) + Q_C$$

Number of Stages at Low or Very High Concentrations

At very low concentrations, it is very difficult to draw the stages even if the scales are enlarged. In such a case, the equilibrium data and the operating line in the low concentration region are plotted on log-log paper. Construction of the stages is carried out with higher accuracy compared to the normal equilibrium diagram because the operating lines and the equilibrium lines do not converge rapidly. At low concentrations, Henry's law applies and $y = kx$. This relation is used to plot the equilibrium line if data are not available. The number of theoretical plates in the low concentration region can also be calculated using Smoker's analytical solution.

REFERENCES

1. Das, D.K., and R.K. Prabhudesai, *Chemical Engineering License Review*, 2nd ed., Kaplan AEC Education, 2004.
2. Henley, E.J. (ed.), *Binary Distillation*, AIChE Modular Instruction, series B, Vol. 1, 1980.
3. Henley, E.J., and J.M. Calo (eds.), Multicomponent Distillation, *AIChE Modular Instruction, Series B*
4. Hougen, O.A., *Chemical Process Principles, part 11, Thermodynamics*, New York, John Wiley & Sons,
5. O'Connell, H.C., *Transactions American Institute of Chemical Engineers*, 42:751, 1946.

6. Perry, R.H. (ed.), *Chemical Engineers' Hand Book*, Platinum Ed., New York, McGraw-Hill, 1999.
7. Robinson, C.S., and E.R. Gilliland, *Elements of Fractional Distillation*, New York, McGraw-Hill.
8. Smith, J.M., and H.C. Van Ness, *Introduction to Chemical Engineering Thermodynamics*, 6th ed., New York, McGraw-Hill,
9. Treybal, R.E., *Mass Transfer Operations*, New York, McGraw-Hill,
10. Van Winkle, M., *Distillation*, New York, McGraw-Hill.

PROBLEMS

The FE exam will usually include several questions on distillation using the McCabe-Thiele diagram.

The following data applies to Problems 14.1 and 14.2:

At a temperature of 366.4 K, the vapor pressures of n-hexane and octane are 1480 and 278 mmHg, respectively. Assume the hexane-octane system obeys Raoult's law and the total pressure is 1 atm.

14.1 Calculate the equilibrium liquid composition of the more volatile component.

14.2 Calculate the equilibrium vapor composition of the more volatile component.

The following data applies to Problems 14.3 through 14.5:

A mixture of 50 mol% n-pentane and 50 mol% n-hexane is fractionated in a column to produce an overhead product containing 95 mol% pentane and a bottom product containing 5 mol% pentane. The distillation is carried out at 1 atm. The equilibrium diagram is shown in Exhibit 14.3. The feed enters the column at its bubble point. Total condenser is used.

14.3 Calculate the minimum reflux ratio.

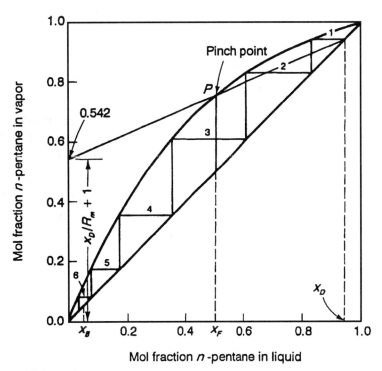

Exhibit 14.3

14.4 Determine minimum number of theoretical stages required. Assume the reboiler is not a theoretical stage.

14.5 Calculate minimum number of theoretical plates using the Fenske equation.

The following data applies to Problems 14.6 through 14.9:

A fractionating column is to be designed to separate 11,000 lb/h of a mixture of 40 mol% benzene and 60 mol% toluene to produce a distillate containing 97 mol% benzene and a bottom product containing 98 mol% toluene. The column is to be operated at one atmosphere and total condenser is to be used. The reboiler is not to be considered as a theoretical stage. The other data are as follows:
Bubble point of feed liquid = 202.4 °F
Feed temperature = 67.4 °F, feed specific heat = 38.2 Btu/(lbmol °F)
Molal latent heat of vaporization for benzene = 13,278 Btu/lbmol
Molal latent heat of vaporization for toluene = 14,355 Btu/lbmol
Equilibrium diagram for benzene-toluene system is given in Exhibit 14.7.

14.6 Calculate the slope of the q line.

14.7 Find the minimum reflux ratio.

14.8 Determine the number of theoretical stages required if a reflux ratio 2 times the minimum is used.

14.9 Determine the feed plate location.

14.10 Find the material balance and the condenser and reboiler duties. Distillate temperature = 179 °F, Bottoms temperature = 229 °F. Neglect heat losses.

14.11 Calculate the liquid and vapor molal overflows in the rectifying and stripping sections.

14.12 The following vapor pressure data were calculated using Antoine constants for the system consisting of 50 mol% benzene, 30 mol% toluene, and 20% ethyl benzene. This system can be assumed to behave ideally.

Vapor Pressure Data System: Benzene, Toluene, and Ethyl Benzene

Component Temperature °C	Benzene mmHg	Toluene mmHg	Ethyl Benzene mmHg
80	758.7	293.0	125.9
85	882.8	347.2	151.8
90	1022.2	409.1	182.0
95	1178.2	479.5	217.0
100	1352.0	559.3	257.1

What is the bubble point temperature of this ternary liquid mixture if the system pressure is 760 mmHg?

SOLUTIONS

14.1 By Raoult's law,

$$p_1 = x_1 P_1 \quad \text{and} \quad p_2 = (1-x_1)P_2$$
$$P_t = P_1 + P_2 = x_1 P_1 + (1-x_1)P_2$$

Hence $x_1 = \dfrac{P_t - P_2}{P_1 - P_2}$

Using the given values of vapor pressure, $x_1 = \dfrac{760-278}{1480-278} = 0.401$

14.2 and $y_1 = \dfrac{x_1 P_1}{P_t} = \dfrac{(0.401)(1480)}{760} = 0.781$

Note:

Similar calculations can be done at other temperatures with corresponding vapor pressures to construct $T\text{-}x\text{-}y$ and $x\text{-}y$ diagrams.

14.3 $x_F = 0.5,\ x_D = 0.95,\ x_B = 0.05$

Locate these points on the equilibrium diagram (Figure 14.5). Feed is at its boiling point. Therefore the q line is vertical. It passes through the intersection of a 45° line and the vertical from $x_F = 0.5$. Draw the q line to intersect the equilibrium line. This point is the pinch point. From (x_D, y_D), draw a line to pass through the pinch point to cut the y-axis to give an intercept = 0.542.

$$\text{intercept} = \dfrac{x_D}{R_m + 1} = 0.542 \text{ Then } R_m + 1 = 0.95/0.542$$

$$R_m = 1.75 - 1 = 0.75$$

Minimum reflux ratio is 0.75

14.4 Minimum number of stages is obtained when the operating lines coincide with a 45° line on the equilibrium diagram. Step off the plates from $x_D = 0.95$. Number of stages is 5.6 (Figure 14.5). According to the problem the reboiler is not to be considered as a theoretical stage.

Then the minimum number of stages = 5.6 as obtained from the equilibrium diagram.

14.5
$$\alpha = \dfrac{y}{1-y}\left(\dfrac{1-x}{x}\right)$$

From Figure 14.5, $x_D = 0.95$, equilibrium composition of vapor = $y_0 = 0.98$.
$x_B = 0.05$ equilibrium composition of vapor = $y_B = 0.11$.

$$\alpha_{\text{top}} = \dfrac{0.98}{1-0.98}\left(\dfrac{1-0.95}{0.95}\right) = 2.58 \qquad \alpha_{\text{bottom}} = \dfrac{0.11}{1-0.11}\left(\dfrac{1-0.5}{0.05}\right) = 2.35$$

$$\alpha_{av} = \sqrt{2.58 \times 2.35} = 2.46$$

$$N_{\min} = \dfrac{\log[0.95(1-0.05)/(0.05(1-0.95))]}{\log 2.46} - 1 = 5.5$$

theoretical stages in the column

14.6 Heat required to raise 1 mol of feed to 202.4 °F = (38.2)(202.4 − 67.4) = 5157 Btu/lbmol

λ of feed mixture = 0.4(13,278) + 0.6(14,355) = 13,924 Btu/lbmol

Therefore, $q = \dfrac{13{,}924 + 5157}{13{,}924} = 1.37$

slope of q-line $= \dfrac{q}{q-1} = \dfrac{1.37}{1.37-1} = 3.7$

14.7 Refer to Exhibit 14.7 for solution.

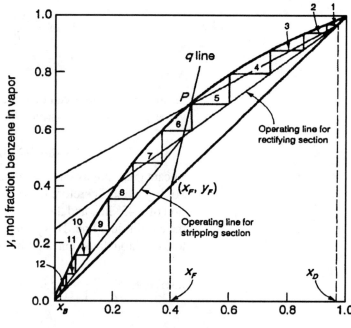

Exhibit 14.7 Equilibrium diagram for benzene-toluene system and solution of Problems 14.7, 8, and 9.

First draw the q line with a slope of 3.7 and passing through the intersection point (x_F, y_F) of vertical from x_F and the 45° line to intersect the equilibrium line at pinch point P. Join (x_D, y_D) and P and produce to meet y-axis at intercept = 0.425.

Thus $\dfrac{x_D}{R_m + 1} = 0.425$, from which $R_m = 1.28$

Minimum reflux ratio is 1.28.

14.8 Actual reflux ratio = 2 × 1.28 = 2.56.

Intercept of rectifying section operating line on y-axis $= \dfrac{x_D}{R+1} = \dfrac{0.97}{2.56+1} = 0.27$.

Operating line for rectifying section is drawn from (x_D, y_D) to intersect y-axis at $y = 0.27$ (Exhibit 14.7).

Draw the operating line for stripping section. Step off the stages in usual manner.

Number of theoretical plates = 12.6.

According to the problem statement, the reboiler is not to be considered as a theoretical stage.

14.9 Crossover from one operating line to the other occurs at the sixth plate. Minimum theoretical stages are obtained when feed is added at the crossover point.

Feed plate = 6.

(Note that this is a theoretical stage. Actual feed tray will depend upon the efficiency of the fractionation.)

14.10 Material balance (overall): $F = D + B$

Component balance: $Fx_F = Dx_D + Bx_B$

Molecular weight of feed = 0.4(78) + 0.6(92) = 86.4

Feed = (11,000/86.4) = 127.3 lbmol $x_D = 0.97$ $x_B = 0.02$

Substitute these values in the material balance equations and get the following

$$D + B = 127.3$$
$$D(0.97) + B(0.02) = 127.3(0.4)$$

Solving, $D = 50.92$ lbmol/h $B = 76.38$ lbmol/h.

Condenser duty

Assume no subcooling.

Vapor flow from the column to the condenser = 2.56 × 50.92 + 50.92 = 181.3 lbmol/h

$\lambda = 0.97(13,278) + 0.03(14,355) = 13,310$ Btu/lbmol of distillate.

Then condenser duty = $Q_c = 181.3 \times 13,310 = 2.413 \times 10^6$ Btu/h

Reboiler duty

Assume molal specific heat of liquid is independent of composition.

Q_{SD} = heat added to distillate = 50.92(38.2)(179 − 67.4) = 217,078 Btu/h

Q_{SB} = heat added to bottoms product = 76.38(38.2)(229 − 67.4) = 471,503 Btu/h

Reboiler duty = $Q_C + Q_{SD} + Q_{SB} = 2.413 \times 10^6 + 217078 + 471503 = 3.102 \times 10^6$ Btu/h

14.11 Flows from plate to plate in rectifying section:

Liquid flow: $L = L_0 = RD = 2.56D = 2.56(50.92) = 130.36$ lbmol/h

Vapor flow: $V = L_0 + D = D(R + 1) = 50.92(2.56 + 1) = 181.28$ lbmol/h

Flows in stripping section:

$$\bar{L} = L + qF = 50.92 \times 2.56 + 1.37 \times 127.3 = 304.76 \text{ lbmol/h}$$
$$\bar{V} = V - (1-q)F = 181.28 - (1-1.37)127.3 = 228.38 \text{ lbmol/h}$$

14.12 Since the system is ideal, the components obey Raoult's law. Hence $p_i = x_i P_i$.

At the bubble point temperature, $\sum_{1}^{n=3} y_i = 1.0$.

Assume a temperature, get vapor pressure for each component, and calculate y_i for each component.

Thus at 82°C for benzene, vapor pressure = 806.6 mmHg and

$$y_{\text{benzene}} = x_{\text{benzene}} P_{\text{benzene}} / P_t = 0.5(758.66)/760 = 0.4991$$
$$y_{\text{toluene}} = x_{\text{toluene}} P_{\text{toluene}} / P_t = 0.3(293.0)/760 = 0.1157$$
$$y_{\text{Ebenzene}} = x_{\text{Ebenzene}} P_{\text{Ebenzene}} / P_t = 0.2(135.8)/760 = 0.0331$$

Then $\sum_{1}^{n=3} y_i = 0.4991 + 0.1157 + 0.0331 = 0.6479 < 1.0$

Therefore a new temperature has to be assumed and calculations repeated.

Similar calculations are done for all components at four different temperatures. The results are given in the following table.

Calculation Results

T°C	y_{benzene}	y_{toluene}	$y_{\text{etylbenzene}}$	Σy_i
80	0.4991	0.1157	0.0331	0.6479
85	0.5808	0.1371	0.0400	0.7578
90	0.6725	0.1615	0.0479	0.8819
95	0.7752	0.1893	0.0571	1.0215

The values of Σy_i are plotted against temperature in Exhibit 14.12.

At $\Sigma y_i = 1.0$, the temperature value is 94.3°C. Therefore, bubble point temperature is 94.3°C.

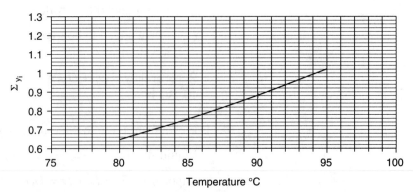

Exhibit 14.12 Graphical solution to finding the bubble point temperature

APPENDIX

Afternoon Sample Examination

INSTRUCTIONS FOR AFTERNOON SESSION

1. You have four hours to work on the afternoon session. You may use the *Fundamentals of Engineering Supplied-Reference Handbook* as your *only* reference. Do not write in the handbook.

2. Answer every question. There is no penalty for guessing.

3. Work rapidly and use your time effectively. If you do not know the correct answer, skip it and return to it later.

4. Some problems are presented in both metric and English units. Solve either problem.

5. Mark your answer sheet carefully. Fill in the answer space completely. No marks on the workbook will be evaluated. Multiple answers receive no credit. If you make a mistake, erase completely.

Work 60 afternoon problems in four hours.

FUNDAMENTALS OF ENGINEERING EXAM

AFTERNOON SESSION

Ⓐ Ⓑ Ⓒ Fill in the circle that matches your exam booklet

1 Ⓐ Ⓑ Ⓒ Ⓓ	16 Ⓐ Ⓑ Ⓒ Ⓓ	31 Ⓐ Ⓑ Ⓒ Ⓓ	46 Ⓐ Ⓑ Ⓒ Ⓓ
2 Ⓐ Ⓑ Ⓒ Ⓓ	17 Ⓐ Ⓑ Ⓒ Ⓓ	32 Ⓐ Ⓑ Ⓒ Ⓓ	47 Ⓐ Ⓑ Ⓒ Ⓓ
3 Ⓐ Ⓑ Ⓒ Ⓓ	18 Ⓐ Ⓑ Ⓒ Ⓓ	33 Ⓐ Ⓑ Ⓒ Ⓓ	48 Ⓐ Ⓑ Ⓒ Ⓓ
4 Ⓐ Ⓑ Ⓒ Ⓓ	19 Ⓐ Ⓑ Ⓒ Ⓓ	34 Ⓐ Ⓑ Ⓒ Ⓓ	49 Ⓐ Ⓑ Ⓒ Ⓓ
5 Ⓐ Ⓑ Ⓒ Ⓓ	20 Ⓐ Ⓑ Ⓒ Ⓓ	35 Ⓐ Ⓑ Ⓒ Ⓓ	50 Ⓐ Ⓑ Ⓒ Ⓓ
6 Ⓐ Ⓑ Ⓒ Ⓓ	21 Ⓐ Ⓑ Ⓒ Ⓓ	36 Ⓐ Ⓑ Ⓒ Ⓓ	51 Ⓐ Ⓑ Ⓒ Ⓓ
7 Ⓐ Ⓑ Ⓒ Ⓓ	22 Ⓐ Ⓑ Ⓒ Ⓓ	37 Ⓐ Ⓑ Ⓒ Ⓓ	52 Ⓐ Ⓑ Ⓒ Ⓓ
8 Ⓐ Ⓑ Ⓒ Ⓓ	23 Ⓐ Ⓑ Ⓒ Ⓓ	38 Ⓐ Ⓑ Ⓒ Ⓓ	53 Ⓐ Ⓑ Ⓒ Ⓓ
9 Ⓐ Ⓑ Ⓒ Ⓓ	24 Ⓐ Ⓑ Ⓒ Ⓓ	39 Ⓐ Ⓑ Ⓒ Ⓓ	54 Ⓐ Ⓑ Ⓒ Ⓓ
10 Ⓐ Ⓑ Ⓒ Ⓓ	25 Ⓐ Ⓑ Ⓒ Ⓓ	40 Ⓐ Ⓑ Ⓒ Ⓓ	55 Ⓐ Ⓑ Ⓒ Ⓓ
11 Ⓐ Ⓑ Ⓒ Ⓓ	26 Ⓐ Ⓑ Ⓒ Ⓓ	41 Ⓐ Ⓑ Ⓒ Ⓓ	56 Ⓐ Ⓑ Ⓒ Ⓓ
12 Ⓐ Ⓑ Ⓒ Ⓓ	27 Ⓐ Ⓑ Ⓒ Ⓓ	42 Ⓐ Ⓑ Ⓒ Ⓓ	57 Ⓐ Ⓑ Ⓒ Ⓓ
13 Ⓐ Ⓑ Ⓒ Ⓓ	28 Ⓐ Ⓑ Ⓒ Ⓓ	43 Ⓐ Ⓑ Ⓒ Ⓓ	58 Ⓐ Ⓑ Ⓒ Ⓓ
14 Ⓐ Ⓑ Ⓒ Ⓓ	29 Ⓐ Ⓑ Ⓒ Ⓓ	44 Ⓐ Ⓑ Ⓒ Ⓓ	59 Ⓐ Ⓑ Ⓒ Ⓓ
15 Ⓐ Ⓑ Ⓒ Ⓓ	30 Ⓐ Ⓑ Ⓒ Ⓓ	45 Ⓐ Ⓑ Ⓒ Ⓓ	60 Ⓐ Ⓑ Ⓒ Ⓓ

DO NOT WRITE IN BLANK AREAS

PROBLEMS

1–3

Iron and steam react according to the following equation:

$$3Fe + 4H_2O \rightarrow Fe_3O_4 + 4H_2$$

Assuming no excess reactants are used and the reaction is complete

1. The amount of iron (kg) required to produce 100 kg of hydrogen is
   a. 168   c. 2100
   b. 2900   d. 1050

2. The amount of iron oxide (kg) produced is
   a. 2320   c. 2100
   b. 2900   d. 1450

3. The volume (m^3) H$_2$ would occupy at standard conditions if it were an ideal gas is
   a. 2240   c. 1184
   b. 1120   d. 17950

4. Vapor pressures of ethyl acetate are 3.22 kPa and 845.8 kPa at 0 and 160 °C, respectively. Its vapor pressure at 100 °C (kPa) is
   a. 183.2   c. 137.9
   b. 1.87   d. 27.1

5–7

10 kmols of the fuel gas (composition by volume: methane 85%, ethane 10.5%, nitrogen 4.5%) is burned with 15% excess air. If complete combustion is assumed, answer the following.

5. Total number of moles after combustion per 10 kmols of gas is
   a. 1236.4   c. 33.9
   b. 272   d. 124

6. If the total pressure is 100.6 kPa, the partial pressure of water vapor (kPa) is
   a. 16.4   c. 0.16
   b. 2.7   d. 2.38

7. A mixture of methyl alcohol and nitrogen at a total pressure of 189 kPa contains 25 vol% methyl alcohol. If the vapor pressure of methyl alcohol can be expressed by a two-constant equation: $\log_{10} p = 7.846 - 1978.4/T$ where p is in kPa and T is in degrees K, the dew point of the mixture (°C) is very nearly
   a. 320   c. 47
   b. 117   d. 50

8. A vessel contains a liquid mixture of 50% benzene and 50% toluene by weight at 100°C. The vapor pressures at 100°C are as follows (assume Raoult's Law is followed):

   Benzene C_6H_6 $p = 1340$ mmHg   Toluene C_6H_8 $p = 560$ mmHg

   The average molecular weight of the vapor in contact with the solution is very nearly
   a. 82
   b. 88
   c. 78
   d. 92

9. A solution of Na_2SO_4 in water is saturated at 50°C. When a saturated solution of Na_2SO_4 is cooled, crystals of $Na_2SO_4 \cdot 10H_2O$ separate from the solution

Temperature, °C	Solubility of Na_2SO_4, g/100g water
50	46.7
10	9

   If 1000 kg of this solution is cooled to 10°C, the percentage yield obtained is nearly
   a. 91
   b. 90
   c. 100
   d. 80

10. A Carnot engine absorbs 1000 kJ at 827°C and rejects heat at 27°C. The work done by the engine is closest to
    a. 1000 kJ
    b. 730 kJ
    c. 550 kJ
    d. 827 kJ

11. How many degrees of freedom does a system of partially decomposed $CaCO_3$ have?
    a. 1
    b. 2
    c. 3
    d. 4

12. What is the molar specific volume of a gas in m³/kg mol at 1.722 MPa and 99°C if the compressibility of the gas at these conditions is 0.87?
    a. 5.1
    b. 2.1
    c. 1.6
    d. 3.5

13. A heat engine absorbs 1055 kJ at 427°C and rejects heat at 38°C. The work done in kJ by the engine if its efficiency is 50% of the Carnot efficiency is closest to
    a. 496.6
    b. 293.3
    c. 1265
    d. 320

14. 80 kg of water at 95°C is adiabatically mixed with 20 kg of cold water at 40°C. The entropy change for the process is closest to
    a. 0.8473 kJ/K
    b. 5.0 kJ/K
    c. −0.8473 kJ/K
    d. −5.0 kJ/K

15. The following data are available for the refrigerant HFC-134a:

T °C	30	40	50
P Bar	7.709	10.174	13.187
v_l m³/kg	8.431×10^{-4}	8.744×10^{-4}	9.056×10^{-4}
v_g m³/kg	0.02665	0.01998	0.01511

    On the basis of the preceding data, the heat of vaporization of HFC-134a at 40 °C (kJ/kg) is closest to
    a. 85 c. 171
    b. 110 d. 152

16. The two-component system of phenol and O-cresol is found to obey Raoult's law. At a temperature of 117 °C, the vapor pressures of phenol and O-cresol are 11.6 (P_1) and 8.76 (P_2) kPa, respectively. The operating total pressure (P_t) is 10 kPa. The relative volatility of component 1 with respect to component 2 is given by

    $$y_1 = \frac{P_1 x_1}{P_t} \quad x_1 = \frac{P_t - P_2}{P_1 - P_2} \quad \alpha_{12} = \left(\frac{y_1}{1-y_1}\right)\left(\frac{1-x_1}{x_1}\right)$$

    The relative volatility of phenol relative to O-cresol at these conditions is closest to
    a. 1.28 c. 1.16
    b. 1.325 d. 1.5

17. A benzene-toluene feed (with 40 mol% benzene and 60 mol% toluene) to a distillation column is at a temperature of 20 °C. The molar heat capacity of the feed is 159.2 kJ/(kg mol·K). Molar latent heats of vaporization of benzene and toluene are 30,813 and 33,325 kJ/kmol, respectively. The bubble point of the mixture is 95 °C. The slope of the q line is closest to
    a. 0 c. 3.7
    b. 1 d. −3.7

18. In absorption of NH_3 by water in a wetted wall column, the mass transfer coefficients in the gas and liquid phases were determined to be 0.6 kg mol/(h·m²·bar) and 40 kg mol/(h·m²·atm), respectively. The equilibrium relationship over the concentration range used in the experiment is given by $y = 0.9x$. The overall mass transfer coefficient based on the liquid phase in kg mol/h·m³ is
    a. 5.33 c. 1.88
    b. 0.533 d. 0.0533

19. The following data pertain to a distillation column fractionating a feed mixture of 40 mol% benzene and 60 mol% toluene.

    Feed temperature = 20 °C (t_F)
    Distillate D: flow = 42 kg mol/h, composition: Benzene 97 mol%
    Bottoms product B: flow = 63 kg mol/h, composition: Toluene 98 mol%
    Bottom product temperature = 109.4 °C (t_B)

Distillate temperature = 82.2 °C (t_D)
External reflux ratio $R = 3.7$
Molar heat capacity of feed bottoms and distillate = 159.2 kJ/(kg mol·K)
Latent heats of vaporization: Benzene 30,813 kJ/kg mol (λ_1)
Toluene 33,325 kJ/kg mol (λ_2)
Enthalpy of distillate relative to feed = $Dc_p(t_D - t_F)$
Enthalpy of bottoms relative to feed = $Bc_p(t_B - t_F)$
Condenser duty = $d(R + 1)\lambda_{AV}$
$$\lambda_{AV} = x_1\lambda_1 + x_2\lambda_2$$

Assume total condenser

The reboiler duty in units of kW is closest to
a. 264   c. 380
b. 2058   d. 1700

20. In absorption of NH_3, the mass transfer coefficient in the gas phase was determined to be 0.59 kg mol/(h·m^2) and that in the liquid phase was found to be 40 kg mol/(h·m^2). The slope of the equilibrium line in the range of experiments was 0.9. The statement that correctly applies to mass transfer in this system is
   a. The liquid phase controls the mass transfer in this system.
   b. The gas phase controls the mass transfer in this system.
   c. Both the gas and liquid resistances are equally significant.
   d. Gas phase controls the mass-transfer, therefore, gas phase resistance is negligible.

21. Acetone is to be absorbed from an air stream containing 3 mol% acetone in an existing packed tower of 1.83 m diameter. The gas flow rate will be 212.5 m^3/min at 30°C and 1.2 bar. The overall mass transfer coefficient for acetone absorption in water may be taken as 230 kg mol/h·m^3 (mol fraction). Then the overall height (m) of a gas transfer unit (HTU) is closest to
   a. 0.8   c. 0.6
   b. 1.0   d. 1.16

22. A reacts at 25 °C according to the following reactions:

$$A + B \xrightarrow{k_1} 2R + S \qquad \Delta H_R = -125.52 \text{ kJ/gmol } A$$
$$A + B \xrightarrow{k_2} P \qquad \Delta H_R = -83.08 \text{ kJ/gmol } A$$

The reaction rate constants for the two reactions are as follows

$$k_1 = 1.2 \frac{\text{liters}}{\text{gmol}\cdot\text{min}} \qquad k_2 = 0.3 \frac{\text{liters}}{\text{gmol}\cdot\text{min}}$$

If 2 moles of A react, the total heat of reaction in kJ is
a. 234.3   c. −251
b. −234.3   d. −209

23. A batch reactor is used to convert a liquid reactant A. The reaction follows second order kinetics. The time in hours required to convert 60% of A if the initial concentration of A is 0.8 gmol/L and $k = 4$ L/(gmol·h) is
    a. 4.7
    b. 0.94
    c. 0.47
    d. 0.5

24. The following data are obtained at 200°C for the reaction $2A \rightarrow 2B + C$. Initially A is pure.

    The concentration of A changes as follows:

Time (min)	Concentration of A gmol/L
0	0.02
200	0.01585
300	0.01435
500	0.0128

    The order of the reaction is
    a. 2
    b. 1
    c. 1.5
    d. 3

25. A homogenous gas phase reaction $A \rightarrow 2R$ has a reaction rate at 250°C and 2 bar as $-r_A = 10^{-2} C_A$ (gmol/liter·s). The space-time in minutes needed for 90% conversion of A in a plug-flow reactor operating at 250°C and 2 bar if the feed to the reactor is 50% A and 50% inerts is
    a. 300
    b. 30
    c. 60
    d. 5

26–27

A liquid phase reaction $A \rightarrow R$ is carried out in a series of three completely mixed stirred tank reactors of equal size. The reaction rate constant k is 0.066 min^{-1}. Overall conversion is 90%. The feed rate is 10 liters/min and the feed contains only A in concentration of 1 gmol/L.

26. The overall space-time in minutes for the three-reactor system is
    a. 5.24
    b. 524
    c. 52.4
    d. 17.5

27. The concentration from the second reactor is
    a. 0.2148
    b. 0.1
    c. 0.35
    d. 0.5

28. The weight of a metal sphere is 1000 N at a place where the local acceleration of gravity g is 10 m/s^2. The weight of the metal sphere on the surface of the moon where the acceleration due to gravity is 1.67 m/s^2 is closest to
    a. 10 N
    b. 170 N
    c. 250 N
    d. 500 N

29. A barometer reads 800 mmHg. If the specific gravity of mercury is 13.6, and g has the value of 9.807 m/s^2, the atmospheric pressure at this location in millibars is closest to
    a. 953 mBar
    b. 990 mBar
    c. 1030 mBar
    d. 1070 mBar

30. For a fractionating column separating a mixture of benzene and toluene, the slope of the q line was –0.5. Twelve theoretical stages are stepped off on the McCabe-Thiele diagram. The operating lines for the rectifying and stripping sections intersect on the q line between the seventh and eighth theoretical stages measured from the top of the column. The overall efficiency of distillation is 70%. Actual feed tray location will be on tray
    a. 7
    b. 8
    c. 10
    d. 11

31. The fixed cost per day of a plant producing a certain type of chemical is $10,000/day. The selling price of the chemical is $200/kg. The variable cost may expressed as $50 + 0.1P$ ($/kg), where P = production rate in kg/day. Find the break-even capacity of the plant.
    a. 70
    b. 1430
    c. 70 or 1430
    d. 100

32. The feed to a binary distillation column is 25 mol% vapor and the balance liquid. The slope of the q line with respect to the positive direction of the x-axis on the x-y diagram is closest to
    a. 0.67
    b. 3.0
    c. –3.0
    d. –0.67

33. The total annual venture cost for insulating an oven may be expressed by

    $$\text{Total annual venture cost} = \frac{870}{0.25 + 2.78T} + 17.25T$$

    where T = insulation thickness. Find the optimum economic insulation thickness.
    a. 2.5
    b. 3.4
    c. 1.3
    d. 4.2

34. The walls of a brick-lined house consists of the following layers of materials:

    Brick layer, 0.1 m thick, $k = 0.8$ W/m·K
    Rock-wool insulation, 0.0762 m thick, $k = 0.065$ W/m·K
    Gypsum plaster board, 0.0375 m thick, $k = 0.5$ W/m·K

    If the inside of the house is maintained at 295 K, estimate the heat loss by conduction through the walls of area 200 m^2 when the outside temperature is 265 K.
    a. 4400 W
    b. 2200 W
    c. 8400 J/s
    d. Insufficient information

35. The overall heat transfer coefficient (U) of a clean shell and tube heat exchanger with negligible tube wall resistance and equal film coefficients is 100 W/m$^2\cdot$K. If the shell side coefficient remains constant, but the tube side coefficient is lowered by 10%, find the ratio $U_{final}/U_{original}$.
    a. 0.80
    b. 0.95
    c. 0.90
    d. 1.10

36. The following information on the terminal temperatures is available for a (2–4) shell-and-tube countercurrent heat exchanger. Estimate the corrected LMTD if the LMTD correction factor is 0.98.

	Shell Side	Tube Side
Inlet temperature (K)	500	200
Outlet temperature (K)	300	290

    a. 148.261
    b. 334.550
    c. 145.300
    d. 83.550

37. A batch is to be cooled from 5 °C to –17 °C in a stirred tank jacketed vessel with an isothermal cooling medium vaporizing at –26 °C. The jacket has a heat transfer surface of 10 m^2 with an overall heat transfer coefficient of 850 W/m$^2\cdot$K. What will be the peak instantaneous cooling load in W, ignoring the heat input from the stirrer and assuming the jacket area is fully wetted by the batch?
    a. 120,000
    b. 1,120,000
    c. 163,500
    d. 263,500

38. A double-pane window assembly consists of two glass plates (0.0025 m thick (k_{glass} = 0.79 W/m$\cdot$K) with 0.01 m air gap between the plates (k_{air} = 0.026 W/m$\cdot$K). The temperature difference between the inside and outside is 25 K. The inside and outside convection coefficients are 6 W/m$^2\cdot$K and 25 W/m$^2\cdot$K respectively. The heat loss through a double-pane glass window in W/m^2 is closest to
    a. 20 W/m^2
    b. 40 W/m^2
    c. 60 W/m^2
    d. 80 W/m^2

39. An existing cocurrent double-pipe heat exchanger cools an organic liquid stream with cooling water. The following data are available:

    Inlet temperature of organics = 500 °F
    Outlet temperature of organics = 400 °F
    Inlet temperature of cooling water = 80 °F
    Outlet temperature of cooling water = 100 °F

    It is required to lower the outlet temperature of the organics to 350 °F from the current temperature of 400 °F by increasing the length of the exchanger. Assuming the flow rates, average heat capacities of the fluids, and overall heat transfer coefficient remain constant, what is the ratio of new length to the original length required?
    a. 1.25
    b. 1.45
    c. 1.66
    d. 1.85

40. A Newtonian liquid is flowing in a pipe at a Reynolds number of 1500. If the viscosity of the liquid is lowered to 50% and density to 60% of their original values by heating, what will be the effect of a pressure drop for the same volumetric flow rate and pipe dimensions?
   a. 50% of original
   b. 60% of original
   c. 25% of original
   d. 30% of original

41. Water is flowing under fully turbulent flow in a pipe. If the diameter of the pipe is lowered by 10%, what will be the effect on pressure drop for the same flow rate?
   a. Increase by 10%
   b. Increase by 21%
   c. Increase by 59.05%
   d. Increase by 69.35%

42. Temperature rise in a heat sensitive organic liquid due to skin friction is to be limited to 10 K. What is the maximum allowable frictional pressure drop if the heat capacity and density of the liquid are 2000 J/kg·K and 500 kg/m^3, respectively?
   a. 1E4 Pa
   b. 1E7 Pa
   c. 1E5 Pa
   d. 1E2 Pa

43. Estimate an approximate pressure drop in psi/100ft for 500 SCFM of air compressed to 100 psig at 60°F flowing through a 3.068-in ID pipe with roughness factor of 0.00015 ft.
   a. 0.09
   b. 0.19
   c. 0.29
   d. 0.39

44. Given: The density of fluid is 1.1×10^3 kg/m^3. The barometric pressure is 100 kN/m^2. The local acceleration due to gravity is 9.81 m/s^2. The absolute static pressure at the bottom of a column of fluid of height 6 m, open to the atmosphere is closest to
   a. 150 kN/m^2
   b. 165 kN/m^2
   c. 65 kN/m^2
   d. 70 kN/m^2

45. In an effort to increase the flow rate of a Newtonian liquid, the supply pressure is increased by 40% in a pipe. Assuming fully turbulent flow before and after the increase of pressure, and no change in density or viscosity, what will be the % increase in flow?
   a. 12
   b. 18
   c. 24
   d. 30

46. A centrifugal fan with a blade diameter of 0.8 m operates at 20 rps to deliver 7.3 m^3/s of air. What would be the capacity in m^3/s of a geometrically similar fan with a 1.23-m impeller blade operating at 30 rps?
   a. 50
   b. 19
   c. 38
   d. 25

47. Feed, distillate, and bottoms product flows and their compositions for a fractionating column are given in the following table (the components are listed in the decreasing order of their K values):

Component	Feed mol %	Distillate mol %	Bottoms Product mol %
Benzene	15.89	18.69	0.0
Toluene	68.79	80.89	0.04
Ethyl benzene	0.7	0.04	4.44
p-Xylene	11.63	0.34	75.77
m-Xylene	1.41	0.03	9.25
o-Xylene	0.31		2.08
Ethyl toluene	0.98		6.55
Heavies	0.28		1.87
Flow (kmol/h)	2222.7	1939.05	283.64

Based on the above data, the heavy key component is
a. Toluene
b. Ethyl benzene
c. p-Xylene
d. Ethyl toluene

48. A shell-and-tube condenser with saturated fluid condensing in the shell side and liquid flowing through the tube side in turbulent flow without change in phase with negligible tube wall resistance has 1000 W/m$^2 \cdot$K film coefficient on the shell side and 125 W/m$^2 \cdot$K on the tube side. Estimate the overall coefficient if the tube side velocity is doubled without changing other physical properties.
a. 179
b. 150
c. 250
d. 161.5

49. What is the upper flammability limit and lower flammability of a mixture of gases whose composition and individual UFL and LFL are given below?

Component	Volume %	LFL %	UFL %
heptane	1	1.1	6.7
hexane	1	1.1	7.5
hydrogen	3	4	75
isopropyl alcohol	1	2	12
Total combustible	6		
air	94		

a. LFL = 1.95%, UFL = 14.7%
b. LFL = 2.7%, UFL = 41.8%
c. LFL = 5.3%, UFL = 28.9%
d. LFL = 2.8%, UFL = 37.4%

50. The time to reduce the concentration of a toxicant from 2.5% by volume to 1 ppm by volume from a vessel of 5000 m^3 in capacity by sweeping an inert at the rate of 3000 m^3/min through the vessel is closest to (assume a nonideal mixing factor of 0.2)
    a. 17 min
    b. 25 min
    c. 85 min
    d. 50 min

51. For the situation given in Problem 50, the number of changes of container-volumes for the duty is closest to
    a. 10
    b. 20
    c. 40
    d. 51

52. Calculate the pH of an aqueous waste containing HCl (0.01 molar) and KCl (0.09 molar) at 25°C.
    a. 2.5
    b. 1.0
    c. 3.0
    d. 2.1

53. 100 kg of saturated liquid water at 350 kPa contained in a closed vessel is heated until 80% of water is vaporized. The amount of heat added to the system is closest to
    a. 58,100 kJ
    b. 93,000 kJ
    c. 127,900 kJ
    d. 172,000 kJ

54. The stack analysis of an ammonia scrubber shows ammonia concentration of 180 ppm (v/v) at 30°C and 1.064 bar. The concentration of ammonia in the stack in mg/m^3 is closest to
    a. 90 mg/m^3
    b. 130 mg/m^3
    c. 170 mg/m^3
    d. 260 mg/m^3

55. Cyanuric chloride has a flash point higher than 374°F. A 75,000-gallon storage tank to store cyanuric chloride liquid is installed so that the tank is not within the diked area or drainage path of Class I or Class II liquids. Is it necessary to size the emergency relief venting of the tank for fire exposure according to NFPA?
    a. yes
    b. no

56. Which of the following substance(s) is (are) not considered a VOC?
    a. Carbon dioxide
    b. Carbon disulfide
    c. Carbon tetrachloride
    d. Benzene

57. Which of the following emission(s) is (are) not a fugitive emission(s)?
    a. Vapor emission from the top vent of a scrubber
    b. Emission from a pump seal
    c. Leakage from a relief valve
    d. Leakage from a compressor

58. The following polynomial has three roots:

$$P(x) = x^3 + 4x^2 + 6x + 4$$

Is $x = -1$ one of the exact roots?
a. No, it is less than the last coefficient
b. Yes, there is a remainder when using synthetic division
c. No, there is a remainder when using synthetic division
d. Yes, there is no remainder

59. For the preceding problem, if a second trial root is to be found (from Equation 11-4) what is this next trial division?
a. 1
b. 2
c. −1
d. −2

60. For a differential equation: $dx/dt + 5x = 1$, $x(0) = 0$; What is the value of x suggested by the Euler's method after the first iteration after the initial condition?
a. 0.1
b. 0.2
c. 0.3
d. 0.4

SOLUTIONS

After each solution, a chapter and chapter section are indicated where you can review the concepts relevant to the problem. If your solution to a problem is incorrect, you may want to review the referenced material.

1–3

Atomic and molecular weights $Fe = 56$, $H_2 = 2$, $H_2O = 18$, $Fe_3O_4 = 232$

Stoichiometric relation $3Fe + 4 H_2O \rightarrow Fe_3O_4 + 4 H_2$

                          168   72   232   8

Basis: 100 kg hydrogen produced

1. **c.** Fe required = $(168/8)(100) = 2100$ kg

    (Chapter 2; "Mass Balances")

2. **b.** Fe_3O_4 produced = $(232/8)(100) = 2900$ kg

    (Chapter 2; "Mass Balances")

3. **b.** $H_2 = 100/2 = 50$ kmol

    1 kmol of an ideal gas occupies 22.4 m³ volume at standard conditions. Therefore, volume of 50 kmols of $H_2 = 50 \times 22.4 = 1120$ m³ at standard conditions

    (Chapter 2; "Ideal Gas Mixtures")

4. **a.** Two data points are available or given. Therefore, use a two-constant equation to correlate the vapor pressure data.

    The two-constant equation is $\ln p = A + B/T$, p in kPa, and T in K

    VP at $0°C = 3.22$ kPa, VP at $160°C = 845.8$ kPa

    By substitution, get two equations as follows:

    $$\ln 3.22 = A + B/(273 + 0) = A + B/273$$
    $$\ln 845.8 = A + B/(273 + 160) = A + B/433$$

    By simplification

    $$273A + B = 319.24 \quad \text{and} \quad 433A + B = 2918.54$$

    Solving simultaneously, $A = 16.246$ and $B = -4116$

    Vapor pressure at 100°C is given by

    $$\ln p = 16.246 - 4116/(273 + 100)$$
    $$\ln p = 5.211, \text{ therefore, } p = 183.3 \text{ kPa}$$

    (Chapter 2; "Vapor Pressure")

5–6

Basis 100 kmols of gas

Reactions are

$$CH_4 + 2O_2 = CO_2 + 2H_2O$$
$$C_2H_6 + 3.5O_2 = 2CO_2 + 3H_2O$$

Assume air composition O_2 = 21% by vol, N_2 = 79% by vol

Prepare a table showing reactants and products as follows:

Component	kmols feed	O_2 consumed kmols (theoretical)	O_2 kmol (Excess)	N_2 kmol in air	products kmol/100 kmol
CH_4	85.0	170			
C_2H_6	10.5	36.5			
N_2	4.5			893.36	897.86
O_2		30.975		30.975	
CO_2					106
H_2O					201.5

5. d. Products per 10 kg mol of fuel gas burned

$$N_2 = 89.79 \text{ kg mol} \quad O_2 = 3.10 \text{ kg mol}$$
$$CO_2 = 10.6 \text{ kg mol} \quad H_2O = 20.15 \text{ kg mol}$$

Therefore, total kg mols/10 kg mol of gas burned = 123.64 kg mol

(Chapter 2; "Mass Balances," "Fuels and Combustion")

6. a. Water vapor in the gas/10 kg mol = 20.15 kg mol

Mol fraction of water = $20.15/123.64 \cong 0.163$

Partial pressure of water vapor in product gas = $0.163 \times 100.6 = 16.4$ kPa

(Chapter 2; "Ideal Gas Mixtures")

7. c. Dew point is the temperature at which partial pressure of methyl alcohol in the mixture will be equal to its vapor pressure at the same temperature.

Total pressure = 189 kPa
Mole fraction of MeOH = 0.25 (volume fraction = mole fraction, ideal behavior assumed)
Partial pressure of MeOH = $0.25 \times 189 = 47.25$ kPa

Substituting in the vapor pressure equation gives

$$\log 47.25 = 7.846 - 1978.4/T$$
$$1.6744 = 7.846 - 1978.4/T$$
$$-6.1716 = -1978.4/T \text{ or } T = \frac{-1978.4}{-6.1716} = 320.6 \text{ K}$$

Therefore, $t = 320.6 - 273 = 47.6 °C$

(Chapter 2; "Vapor Pressure," "Ideal Gas Mixtures")

8. **a.** Basis 100 kg of mixture. Prepare a table as follows:

Component	wt%	wt. kg	MW	kg-mols	x_i (liquid)	P_i kPa	p_i (vapor)	y_i
Benzene	50	50	78	0.6410	0.5412	178.6	96.7	0.7387
Toluene	50	50	92	0.5435	0.4588	74.6	34.2	0.2613
				1.1845	1.0000		982.1	1.0000

Therefore, vapor composition: benzene 73.87 mol% and toluene 26.13 mol%

Average molecular weight of vapor = 0.7387(78) + 0.2613(92) = 81.66

(Chapter 2; "Definitions and Relationships of Material Properties")

9. **a.** At 50°C, solubility of Na_2SO_4 = 46.7 g/100 g of water, i.e., 31.83% by weight

At 10°C, solubility of Na_2SO_4 = 9 g/100 g of water, i.e., 8.26% by weight

Let x be the amount of $Na2SO_4 \cdot 10H_2O$ formed in kg

By material balance on Na_2SO_4

$$1000(0.3183) = (142/322)(x) + (1000 - x)(0.0826)$$

which, on solution, gives

$$x = 657.6 \text{ kg of } Na_2SO_4 \; 10H_2O$$

Na_2SO_4 in original solution = 318.3 kg and Na_2SO_4 in crystals
$$= (0.441)(657.6) = 290 \text{ kg}$$
% yield = (290/318.3) × 100 = 91.1 %

(Chapter 2; "Mass Balances," "Solubility and Crystallization")

10. **b.**
$$T_H = 827 + 273 = 1100 \text{ K}$$
$$T_C = 27 + 273 = 300 \text{ K}$$
Work done = heat absorbed × (Carnot efficiency)
$$= 1000 \times \frac{T_H - T_C}{T_H} = 1000 \times \frac{1100 - 300}{1100} = 727.3 \text{ kJ}$$

(Chapter 3; "Second Law of Thermodynamics")

11. **a.** The decomposition of $CaCO_3$ can be represented by the equation

$$CaCO_3(s) \rightarrow CaO(s) + CO_2(g)$$

Number of components = 3
Number of reactions = 1
Number of phases = 3 (2 solid and 1 gas)
By phase rule, the degrees of freedom = $N - r + 2 - P = 3 - 1 + 2 - 3 = 1$

(Chapter 3; "Gibbs' Phase Rule")

12. c. $$v = \frac{ZRT}{P} = \frac{0.87 \times 0.0008314 \times 372}{1.722} = 1.563 \text{ m}^3/\text{kg mol}$$

    (Chapter 3; "Compressibility Factor")

13. b. $$T_1 = 427 + 273 = 700 \text{ K} \quad T_2 = 38 + 273 = 311 \text{ K}$$

    Carnot efficiency = (700 − 311)/700 = 0.556
    Actual efficiency = 0.556 × 0.5 = 0.278
    Work done = 0.278 × 1055 = 293.3 kJ.

    (Chapter 3; "Second Law of Thermodynamics")

14. a. Assume heat capacity of water constant and equal to 4.184 kJ/kg·K

    By energy balance, 4.184 × 80 (95 − t) = 20(t − 40) × 4.184

    Solving for t gives t = (95 × 80 + 20 × 40)/100 = 84°C = 357 K

    For hot water, $\Delta S_H = 80 \int_{368}^{357} \frac{C_p dT}{T} = 80(4.184)\ln\frac{T_2}{T_1} = 80(4.184)\ln\frac{357}{368}$

    $$= -10.16 \text{ kJ/K}$$

    For cold water, $\Delta S_C = 20 \int_{313}^{357} \frac{C_p dT}{T} = 20(4.184)\ln\frac{357}{313} = +11.007 \text{ kJ/K}$

    Total entropy change for the mixing process is

    $$\Delta S_H + \Delta S_C = -10.16 + 11.00713 = +0.8473 \text{ kJ/K}$$

    (Chapter 3; "Second Law of Thermodynamics")

15. c. Assume ΔH_V = constant over the small temperature range

    Then $\frac{\Delta P}{\Delta T} = \frac{\Delta H}{T \Delta V}$ bar/K

    $\Delta H_V = T (\Delta V)(\Delta P/\Delta T) = (40 + 273)(0.01998 - 8.744 \times 10^{-4})(0.2739)$

    $= 1.7074 \text{ m}^3 \text{ bar/kg}$

    $= 1.7074 \text{ (m}^3 \text{ bar/kg)}/(10^{-2} \text{m}^3 \text{ bar/kJ)}$

    $= 170.74 \text{ kJ/kg}$

    (Chapter 3; "Clapeyron Equation")

16. b. Let x_p = mole fraction phenol in liquid y_p = mole fraction phenol in vapor
    Likewise x_{oc} and y_{oc} are liquid and vapor compositions of O-cresol
    P_p and P_{oc} are vapor pressures of pure phenol and O-cresol, respectively

    By Raoult's law, $y_p = \frac{x_p P_p}{P_T} = \frac{x_p P_p}{10} \quad y_{oc} = \frac{x_{oc} P_{oc}}{P_T} = \frac{x_{oc} P_{oc}}{10}$

    Also $x_{oc} = 1 - x_p$

    By Dalton's relation, $x_p P_p + (1-x_p)P_{oc} = 10$

    from which $x_p = \frac{10 - P_{oc}}{P_p - P_{oc}}$

For given data of the problem

$$x_p = \frac{10 - 8.76}{11.6 - 8.76} = 0.43$$

$$y_p = 0.437 \times 11.6/10 = 0.507$$

$$a_{phenol} = \frac{y_p x_{oc}}{y_{oc} x_p} = \frac{y_p(1-x_p)}{(1-y_p)(x_p)}$$

$$= \frac{0.507(1-0.437)}{(1-0.507)(0.437)} = 1.32$$

(Chapter 14; "Ideal Systems;" "Relative Volatility")

17. **c.** Latent heat of vaporization of the mixture = 0.4(30813) + 0.6(33325) = 32320 kJ/kg mol of feed

Sensible heat to raise the temperature of feed to bubble point = 159.2 (95 − 20) = 11940 kJ/kg mol

Then $q = \dfrac{32320 + 11940}{32320} = 1.37$

and the slope of the q line $= \dfrac{1.37}{1.37 - 1} = 3.7$

(Chapter 14; "Design of Columns for Simple Binary Systems," "McCabe-Thiele Method")

18. **b.** Based on the liquid phase, the overall mass transfer coefficient is given by

$$\frac{1}{K_x} = \frac{1}{k_x} + \frac{1}{mky} = \frac{1}{40} + \frac{1}{0.9 \times (0.6)} = 1.8769$$

Therefore, $K_x = (1/1.6769) = 0.533$ kg mol/h·m²

(Chapter 4; "Mass Transfer Coefficients")

19. **b.** Heat balance on the column gives $Q_R = Q_C + Q_{SB} + Q_{SD}$

Assume specific heat capacities of feed, distillate, and bottoms product same.

$$Q_{SD} = 159.2(82.2 - 20)(42) = 415894 \text{ kJ/h}$$
$$Q_{SB} = 159.2(109.4 - 20)(63) = 896646 \text{ kJ/h}$$

Calculation of Q_C:

Vapor to condenser = $D(R + 1) = 42(3.7 + 1) = 197.4$ kg mol/h
$\lambda = 0.97(30813) + 0.03(33325) = 30,888$ kJ/kg mol
$Q_C = 30,888 \times 197.4 = 6,097,362$ kJ/h

Then the reboiler duty $Q_B = 415,894 + 896,646 + 6,097,362$
$= 7,409,902$ kJ/h
$= 7,409,902/3600 = 2058$ kJ/s $= 2058$ kW

(Chapter 14; "Design of Columns for Simple Binary Systems," "McCabe-Thiele Method")

20. **b.** Calculate the resistances in each phase based on gas phase.

$$\frac{1}{K_y} = \frac{1}{k_y} + \frac{m}{k_x} = \frac{1}{0.59} + \frac{0.9}{40} = 1.695 + 0.0225 = 1.7155$$

The liquid phase resistance is 0.0225 compared to 1.695 in the gas phase.

Therefore, in this system the gas phase controls the mass transfer process.

(Chapter 4; "Overall Mass Transfer Coefficients")

21. **b.** Cross-sectional area of tower = $0.785(1.83)^2 = 2.63 \text{m}^2$

$$\text{Gas rate} = \frac{212.5 \times 60}{22.4} \times \frac{273}{303} \times \frac{1.22}{1.013} = 617.64 \text{ kg-mol/h}$$

$$G = \frac{617.64}{2.63} = 234 \text{ kgmol/h} \cdot \text{m}^2$$

Therefore, $HTU_G = (G/H_y a) = 234/230 = 1.017$ m

(Chapter 4; "Mass Transfer in Packed Beds")

22. **b.**

For the first reaction, $\quad -r_{A1} = k_1 C_A C_B$
For the second reaction, $-r_{A2} = k_2 C_A C_B$

Dividing the first equation by the second,

$$\frac{r_{A1}}{r_{A2}} = \frac{k_1}{k_2} = \frac{1.2}{0.3} = 4.0$$

Thus the ratio of moles of A consumed in reaction 1 to moles of A consumed in the second reaction is 4:1.

Then moles of A reacted in first reaction = 2(4/5) = 1.6 moles.

Also moles of A reacted in second reaction = 2(1/5) = 0.4 moles.

Total heat of reaction = 1.6(–125.52) + 0.4(–83.08) = –234.3/2 moles of A.

(Chapter 5; "Thermodynamics of Chemical Reactions," "Rates of Chemical Reaction")

23. **c.** For a reaction following second order kinetics,

$$-r_A = -\frac{dC_A}{dt} = kC_A^2$$

In terms of conversion X_A, the rate equation becomes

$$-r_A = C_{AO} \frac{dX_A}{dt} = kC_{AO}^2 (1 - X_A)^2$$

Then

$$C_{AO}k\,dt = \int_0^{X_A} \frac{dX_A}{(1-X_A)^2}$$

Solution of the equation is $C_{AO}kt = \dfrac{X_A}{1-X_A}$

For a batch reactor, $t = \dfrac{X_{A'}(1-X_A)}{C_{AO}k} = \dfrac{0.6/(1-0.6)}{0.8 \times 4} = 0.47$ hr

(Chapter 5; "Batch Reactors")

24. **a.** First assume the reaction is second order. The rate equation is

$$-\frac{dC_A}{dt} = kC_A^2 \quad \text{or} \quad -\frac{dC_A}{C_A^2} = k\,dt$$

which on integration with boundary condition $C_A = C_{AO}$ at $t = 0$ yields the solution

$$\frac{1}{C_A} - \frac{1}{C_{AO}} = kt$$

Thus a plot of $\dfrac{1}{C_A}$ versus t will be a straight line. To solve the problem without a graph, calculate k values for each run. If k values thus calculated are the same, the assumed order of the reaction is correct. The calculations are given in the following table.

Time (min)	C_A	$\dfrac{1}{C_A}$	$\dfrac{1}{C_A} - \dfrac{1}{C_{AO}}$	$k = \left(\dfrac{1}{C_A} - \dfrac{1}{C_{AO}}\right)/t$
0	0.02	50.00		
200	0.01585	63.09	13.09	0.0655
300	0.01435	69.68	19.68	0.0656
500	0.01218	82.78	32.78	0.0656

The k values calculated are equal; therefore, assumed order of the reaction is correct

(Chapter 5; "Rate of Chemical Reaction")

25. **d.**

	Change of volume	Initial moles	Final moles
A		1	0
R		0	2
I		1	1
		2	3

Therefore, $\varepsilon_A = \dfrac{3-2}{2} = 0.5$

And the relation between conversion and concentration is

$$\dfrac{C_A}{C_{AO}} = \dfrac{1-X_A}{1+\varepsilon_A X_A} = \dfrac{1-X_A}{1+0.5 X_A}$$

For a plug-flow reactor,

$$\tau = C_{AO}\int_0^{X_A}\dfrac{dX_A}{-r_A} = C_{AO}\int_0^{X_A}\dfrac{dX_A}{10^{-2}C_A}$$

$$= C_{AO}\int_0^{X_A}\dfrac{dX_A}{10^{-2}C_{Ao}(1-X_A)'(1+0.5X_A)}$$

The solution of the preceding equation is

$$10^{-2}\tau = -(1+\varepsilon_A)\ln(1-X_A) - \varepsilon_A X_A$$
$$= -1.5\ln(1-X_A) - 0.5\,X_A$$
$$\tau = 10^2\,[-1.5\ln(1-0.9) - 0.5 \times 0.9]$$
$$= 300.4 \text{ s} = 5 \text{ min.}$$

(Chapter 5; "Steady State Tubular (Plug Flow) Reactors")

26. **c.**

For a system of three reactors, $\tau_n = 3\tau = \dfrac{n}{k}\left[\left(\dfrac{C_{AO}}{C_{An}}\right)^{\frac{1}{n}} - 1\right]$

$$= \dfrac{3}{0.066}\left[\left(\dfrac{1}{0.1}\right)^{\frac{1}{3}} - 1\right]$$

$$= 52.4 \text{ min}$$

(Chapter 5; "Mixed Flow Reactors in Series")

27. **a.** Concentrations of A in exit stream from each reactor, because all reactors are of equal size,

for first order reaction $\dfrac{C_{Ai}-1}{C_{Ai}} = 1 + k\tau$

for each reactor $\tau = 52.4/3 \cong 17.5$ min

then

$$C_{A1} = \dfrac{C_{Ai}-1}{1+k\tau} = 1/(1+0.66\times 17.5) = 0.4635$$
$$C_{A2} = 0.4635(1+0.066\times 17.5) = 0.2148$$
$$C_{A3} = 0.2148(1+(0.066\times 17.5) = 0.1$$

(Chapter 5; "Mixed Flow Reactors in Series")

28. **b.** By Newton's law, $F = ma = mg$ since $a = g$ or since weight and force are the same, weight $= mg$.

Hence, $m = \text{weight}/g = 1000 \text{ N}/(10.0 \text{ m/s}^2) = 100.0 \text{ Nm/s}^2 = 100(1 \text{ kg/m-s}^2) = 100 \text{ kg}$.

Mass of the metal sphere will remain the same regardless of its location. However, its weight will change because of a change in acceleration due to gravity. Since the weight and the gravitational force will be the same on the surface of the moon, one can write weight $= F_{moon} = mg = 100 \text{ kg} \times 1.67 \text{ m/s}^2 = 167 \text{ N}$

(Chapter 1; "Units of Force")

29. **d.** It can be shown that the following relation holds for a column of mercury in a barometric tube.

$$\Delta P = P_A - P_V = -\rho g \Delta Z$$

where
P_A = atmospheric pressure
P_V = Pressure exerted on the surface of the liquid column in the barometric tube
    = 0 for a mercury barometer
Z = height of mercury column in the barometric tube

For mercury, the specific gravity is 13.6, hence its density is $\rho = 13.6 \text{ g/cm}^3 = 13.6 \times 10^3 \text{ kg/cm}^3$.

$$\Delta Z = 800 \text{ mm}/(1000 \text{ mm}) = 0.8 \text{ m}$$
$$P_A = (13.6 \times 10^3 \text{ kg/m}^3) \times (9.807 \text{ m/s}^2) \times (0.8 \text{m})$$
$$= 1.067 \times 10^5 \frac{\text{kg}}{\text{ms}^2}$$
$$= 1.067 \times 10^5 \text{ Pa since } 1 \text{ Pa} = 1\frac{\text{kg}}{\text{ms}^2}$$
$$= 1.067 \times 10^5 \text{ Pa}/(10^5 \text{ Pa/bar})$$
$$= 1.067 \text{ Bar} = 1067 \text{ mbar}$$

A simpler and quicker way to solve this problem is to use the conversion factors from the table.

(Chapter 1)

30. **c.** Number of theoretical stages in the column = 11.

Feed is on 7th theoretical stage measured from the top.

Or 7/0.7 = 10th actual tray in the column.

(Chapter 14; "Column Efficiency," "Feed Plate Location")

31. **c.**

$$Q_B = \frac{F}{s-c}$$

$F = \$10{,}000/\text{day}$
$s = \$200/\text{kg}$
$c = 50 + 0.1P$. At break-even point, $P = $ break-even capacity $= Q_B$ kg/day.

Therefore, at break-even capacity, $c = 50 + 0.1\,Q_B$.

Substituting in the above equation:

$$Q_B = \frac{10000}{200-(50+0.1Q_B)} = \frac{10000}{150-0.1Q_B}$$
$$Q_B^2 = 1500Q_B + 100000 = 0$$

Solving for Q_B, $Q_B = 70$ or 1430

(Chapter 6; "Linear Break-Even Analysis")

32. **c.** Since vapor is 25 mol%, the liquid fraction of the feed is 75 mol%.

Therefore the slope of the q line $= \dfrac{0.75}{0.75-1} = -3.0$

(Chapter 14; "q Values and q-Line Position")

33. **d.**

$$V_c = \frac{870}{0.25+2.78T} + 17.25T$$
$$\frac{dV_c}{dT} = -\frac{870 \times 2.78}{(0.25+2.78T)^2} + 17.25 = 0$$

Hence, $T = 4.2$

(This is a calculus problem)

34. **a.** From equation (7-3):

$$q_k = \frac{T_0 - T_3}{X_1/k_1 A + X_2/k_2 A + X_3/k_3 A}$$
$$= \frac{A(T_0 - T_3)}{X_1/k_1 + X_2/k_2 + X_3/k_3}$$
$$= \frac{200(295-265)}{0.1/0.8 + 0.0762/0.065 + 0.0375/0.5}$$
$$= 4372.2 \text{ W}$$

(Chapter 7; "Conduction")

35. **b.** Originally

$$U_{o1} = 100 = h_o h_i/(h_o + h_i) = h_o/2, \text{ since } h_o = h_i$$
$$h_o = h_i = 2U_{o1} = 200$$

Finally

$$h_o = 200, \ h_i = 0.9 \times 200 = 180$$
$$U_{o2} = 200 \times 180/(200 + 180) = 94.74$$
$$U_{o2}/U_{o1} = 0.9474$$

(Chapter 7; "Fluid to Fluid Heat Transfer Across a Solid Wall")

36. **c.**

$$T_1(500) \text{------------>} T_2(300)$$
$$t_2(290) \text{<------------} t_1(200)$$

$$\Delta T_1 = T_1 - t_2 = 500 - 290 = 210 \text{ K}$$
$$\Delta T_2 = T_2 - t_1 = 300 - 200 = 100 \text{ K}$$
$$\text{LMTD} = \frac{\Delta T_1 - \Delta T_2}{\ln\left[\frac{\Delta T_1}{\Delta T_2}\right]} = \frac{210 - 100}{\ln\left[\frac{210}{100}\right]} = 148.26 \text{ K}$$

Corrected LMTD = $0.98 \times 148.26 = 145.30$ K

(Chapter 7; "Log Mean Temperature Difference (LMTD) and Correction Factor (F_T)")

37. **d.** From Equation (7-37):

$$-Q_{\Theta=0} = UA(T_1 - t_1) = 850 \times 10[5 - (-26)] = 263,000 \text{ W}$$

(Chapter 7; "Isothermal Cooling Medium")

38. **b.** From Equation (7-6a)

$$1/U = 1/h_i + \Sigma X_i/k_i A_i = 1/h_o$$
$$= 1/6 + 2 \times 0.0025/0.79 \times 1 + 0.01/0.026 \times 1 + 1/25$$

2 glass panes air gap

$$= 0.5976 \text{ m}^2 \cdot \text{K/W}$$
$$U = 1.6733 \text{ W/m}^2\text{K}$$

From Equation (7-6b)

$$q = UA\Delta T = 1.6733 \times 1 \times 25 = 41.8325 \text{ W/m}^2$$

(Chapter 7; "Convection")

39. **c.** Existing Design:

$$T_{1e} = 500 \text{ -------------} \rightarrow T_{2e} = 400$$
$$t_{1e} = 80 \text{ ----------------} \rightarrow t_{2e} = 100$$

New design:

$$T_{1e} = 500 \text{ -------------} \rightarrow T_{2n} = 350$$
$$t_{1e} = 80 \text{ ----------------} \rightarrow t_{2n} = ?$$

Subscript n indicates value for the new design.

Calculation:

Assume:
- W = hot fluid mass flow rate
- C = hot fluid heat capacity
- w = cold fluid mass flow rate
- c = cold fluid heat capacity
- U = overall heat transfer coefficient
- A_e = existing heat transfer surface area
- A_n = new heat transfer surface area
- $LMTD_e$ = LMTD for existing design
- $LMTD_n$ = LMTD for new design

From heat balance between fluids for existing operation:

$$R = \frac{wc}{WC} = \frac{T_{1e} - T_{2e}}{t_{2e} - t_{1e}} = \frac{500 - 400}{100 - 80} = 5$$

$$LMTD_e = \frac{((T_{1e} - t_{1e}) - (T_{2e} - t_{2e}))}{\ln\left[\frac{T_{1e} - t_{1e}}{T_{2e} - t_{2e}}\right]} = \frac{(500 - 80) - (400 - 100)}{\ln \frac{500 - 80}{400 - 100}}$$

$$= 356.642$$

An overall heat balance for the existing operation using the overall heat transfer coefficient:

$$Q = UA_e LMTD_e = WC(T_{1e} - T_{2e}) \tag{1}$$

A heat balance between two fluids for the new design:

$$WC(T_{1e} - T_{2n}) = wc(t_{2n} - t_{1e})$$

$$t_{2n} = \frac{T_{1e} - T_{2n}}{\frac{wc}{WC}} + t_{1e} = \frac{T_{1e} - T_{2n}}{R} + t_{1e} = \frac{500 - 350}{5} + 80 = 110$$

$$LMTD_n = \frac{((T_{1e} - t_{1e}) - (T_{2n} - t_{2n}))}{\ln\left[\frac{(T_{1e} - t_{1e})}{(T_{2n} - t_{2n})}\right]} = \frac{((500 - 80) - (350 - 110))}{\ln\left[\frac{500 - 80}{350 - 110}\right]}$$

$$= 321.649$$

An overall heat balance based on overall heat transfer coefficient for the new design:

$$UA_n LMTD_n = WC(T_{1e} - T_{2n}) \qquad (2)$$

Dividing (2) by (1):

$$\frac{A_n}{A_e} = \frac{LMTD_e}{LMTD_n} \frac{T_{1e} - T_{2n}}{T_{1e} - T_{2e}} = \left[\frac{356.642}{321.649}\right]\left[\frac{500-350}{500-400}\right] = 1.663$$

Since surface area is proportional to length, the ratio of new length to original length is 1.663.

(Chapter 7; "Estimation of Outlet Temperatures")

40. **a.** Check N_{Re} for turbulent flow. For the same pipe and same volumetric flow rate $N_{Re} \propto \rho/\mu$.

    $N_{Re2} = N_{Re1}(\rho_2 \mu_1 / \rho_1 \mu_2) = 1500 \times 0.6 \times 1/(1 \times 0.5) = 1800 < 2100$. Flow is still laminar.

    From Equation (8-7): $\Delta P_2/\Delta P_1 = \mu_2/\mu_1 = 0.5/1 = 0.5$.

    Therefore, the pressure drop will be reduced by 50%.

    (Chapter 8; "Steady, Incompressible Flow of Fluid in Conduits and Pipes")

41. **d.** From Equation (8-15) $\Delta P \propto 1/d^5$. Therefore, $\Delta P_2/\Delta P_1 = (d_1/d_2)^5 = (1/0.9)^5 = 1.6935$.

    Hence, pressure drop will increase by 69.35%.

    (Chapter 8; "Reynolds Number")

42. **b.** From Equation (8-33)

    $$\Delta P = \Delta T \rho \eta C_p = 10 \text{ K} \times \frac{500 \text{ kg}}{\text{m}^3} \times \frac{\text{N} \cdot \text{m}}{\text{J}} \times \frac{2000 \text{ J}}{\text{kg} \cdot \text{K}}$$
    $$= 1 \times 10^7 \text{ Pa}$$

    (Chapter 8; "Temperature Rise Due to Skin Friction Under Adiabatic Condition")

43. **b.** Assume fully turbulent flow. From Equation (8-16)

    $$f = \frac{1}{16[\log(3.7 \times 3.068/12 \times 0.000150)]^2}$$
    $$= 0.00432$$

    $$\rho_{standard} = PM/RT = 14.7 \times 29/(10.73 \times 520) = 0.0764 \text{ lb/ft}^3$$
    $$\text{Air mass flow rate} = 500 \times 0.0764 \times 60 = 2292 \text{ lb/h}$$
    $$\rho_{flow} = PM/RT = 114.7 \times 29/(10.73 \times 520) = 0.596 \text{ lb/ft}^3$$

From Equation (8-15)

$$\Delta P_{100} = 0.0000134 \times 0.00432 \times 100 \times (2292)^2/(0.596 \times 3.068^5)$$
$$= 0.19 \text{ psi}$$

(Chapter 8; "Reynolds Number," "Friction Factor," "Newtonian Fluid")

44. **b.** From Equation (8-4)

$$P_h = h\rho g/g_c = 6\text{m} \times \frac{1.1 \times 10^3 \text{kg}}{\text{m}^3} \times \frac{9.81\text{m}}{\text{s}^2} \times \frac{\text{s}^2 \cdot \text{N}}{\text{kg.m}}$$
$$= 64.746 \text{ kN/m}^2$$

Absolute pressure at the bottom of the column = 64.746 + 100 = 164.746 kN/m²

(Chapter 8; "Static Pressure, Static Head")

45. **b.** From Equation (8-13) and from the conditions given by the problem

$$\Delta P \propto Q^2$$

Therefore, $Q_2/Q_1 = (\Delta P_2/\Delta P_1)^{0.5} = (1.4)^{0.5} = 1.1832$

Increase in flow = 18.32%.

(Chapter 8; "Reynolds Number")

46. **c.** From Equation (10-11)

When D is constant $(Q_2/Q_1) = N_2/N_1$, and when N is constant for a geometrically similar fan, $Q_2/Q_1 = (Q_2/D_1)^3$.

By the theorem of joint variation, when both N and D vary,

$$Q_2/Q_1 = (N_2/N_1)(D_2/D_1)^3$$
$$Q_2 = 7(30/20)(1.23/0.8)^3 = 38.16 \text{ m}^3/\text{s}$$

(Chapter 10; "Centrifugal Pumps")

47. **b.** By definition, key components are those whose separation is specified in the distillation. In this case, toluene is mostly taken as distillate product while ethylbenzene is removed from the column as bottoms product. Therefore ethylbenzene is the heavy key component.

(Chapter 14; "Fractional Distillation")

48. **a.** By doubling the tube side velocity, the tube side coefficient will be

$$h_i = 125 \times 2^{0.8} = 217.64$$
$$U = \frac{1000 \times 217.64}{1000 + 217.64} = 178.73$$

(Chapter 7, "Isothermal Heating Medium")

49. **a.**

Component	Volume %	LFL %	UFL %	y_i	y_i/LFL_i	y_i/UFL_i
heptane	1	1.1	6.7	0.167*	0.152**	0.025
hexane	1	1.1	7.5	0.167	0.152	0.022
hydrogen	3	4	75	0.5	0.125	0.007
isopropyl alcohol	1	2	12	0.167	0.084	0.014
Total combustible	6			1.001	0.513	0.068
air	94					

*This is obtained as 1/6.
**This is obtained as 0.167/1.1.

$LFL_{mixture} = 1/0.513 = 1.95\ \%$, and $UFL_{mixture} = 1/0.068 = 14.71\%$

Since the combustible volume % (6%) falls in the flammability range (1.95% and 14.71%), the mixture is flammable.

(Chapter 12; "Calculating Flammability Limits")

50. **c.** Time for sweep-through purging is given by

$$t = \frac{V}{KQ}\ln\left(\frac{C_0}{C}\right) = \frac{5000}{0.2\times 3000}\ln\left(\frac{2.5\times 10^{-2}}{1\times 10^{-6}}\right) = 84.39\ \text{minutes}$$

(Chapter 12; "Sweep-Through Purging")

51. **d.** The number of changes of container volumes $= Qt/V = 3000 \times 84.39/5000 = 50.634$

(Chapter 12; "Sweep-Through Purging")

52. **d.** The equations necessary to compute the pH are:

$$\text{pH} = -\log_{10}(\gamma C_{H_3O^+})$$

$$I = \frac{1}{2}\Sigma\left(C_i Z_i^2\right)$$

$$\log_{10}\gamma_{H_3O^+} = \frac{-0.5I^{0.5}}{1+3I^{0.5}}$$

Substituting the numerical values of concentrations and ionic charges

$$I = \frac{1}{2}(0.01\times 1^2 + 0.01\times 1^2 + 0.09\times 1^2 + 0.09\times 1^2) = 0.1$$

$$\log_{10}\gamma_{H_3O^+} = \frac{-0.5I^{0.5}}{1+3I^{0.5}} = -0.0811388$$

$$\gamma_{H_3O^+} = 10^{-0.0811388} = 0.8296$$

$$\text{pH} = -\log_{10}(0.01\times 0.8296) = 2.08$$

(Chapter 13; "pH")

53. **d.** System: 100 kg of saturated liquid water in a closed vessel.

   At saturation both pressure and temperature remain constant.

   Energy balance reduces to $U = Q$

   $P = 350$ kPa $t = 138.3°C$ from steam tables.

   $x_i = 0.2$ and $x_g = 0.8$ by weight

   $H_g = 2729.2$ kJ/kg and $h_1 = 581.6$ kJ/kg from steam tables

   $\Delta U = Q = 0.8(2729.2) + 0.2(581.6) - 1.0(581.6) = 1718.1$ kJ/kg

   Therefore, total heat added = $100(1718.1) = 171810$ kJ

   (Chapter 3; "Energy Balance," First Law of Thermodynamics")

54. **b.**

   $$1\,\text{ppm} = \frac{PM}{0.08314T} = \frac{1.064 \times 17}{0.08314 \times 303} = 0.718 \frac{mg}{m^3}$$
   $$180\,\text{ppm} = 129.2\,\text{mg/m}^3$$

   (Chapter 12; "Threshold Limit Values")

55. **b.** (Chapter 12; "Classification of Liquids for Thermal Hazards Analysis")

56. **a.** (Chapter 13; "Volatile Organic Compounds (VOC)")

57. **a.** (Chapter 13; "Fugitive Emission")

58. **b.** Synthetic division

   ```
       1    4    6    4  |-1   trial root
           -1   -3   -3
       1    3    3   |1   a remainder of 1
   ```

   (Chapter 11; "Root Extraction," "Newton's Method")

59. **d.**

   $$x_{n+1} = x_n - P(x_n)/P'(x_n)$$
   for $x_1 = -1$, $P'(x) = 3x^2 + 8x + 6\big|_{x=-1} = 3 - 8 + 6 = 1$
   $P(x)\big|_{x=-1} = -1 + 4 - 6 + 4 = +1$
   $\therefore x_{n+1} = (-1) - (1/1) = 2$

   (Chapter 11; "Root Extraction," "Newton's Method")

60. **c.** Using equation 11.9a

   $$(x_{k+1}) = x_k - Tax_k + Tf_k \quad \text{for fixed increments.}$$
   $$x_1 = x_0 - Tax_0 + Tf_0 = 0 - 0 + 0.3 \times 1 = 0.3$$

   (Chapter 11; "Numerical Solutions of Differential Equations")

INDEX

A
Accounting principle, 10, 16
Adiabatic process, 33
Analysis
  integral method of, 81
  proximate, 20
  ultimate, 20

B
Balance
  energy, 17, 29
  mass, 16–17, 29
Boiling point, 11
  elevation, 15
Break-even analysis, 95

C
Carnot cycle, 40–41
  reversed, 41
Carnot engine, 40
Carnot principle, 31
Chemical equilibrium, 71–72
Chemical reactions
  classification of, 71
  rate of, 73–74
  thermodynamics of, 70–71
Combustion, 19–20
Compressibility factor, 39
Concentration
  mass, 11
  molar, 11
  volume, 11
Continuous-flow stirred tank reactors (CFSTR), 84
Conversion factors, 2–4
Cost factors, of installation, 95–96
Critical properties, 13
Crystallization, 16
Cyclic processes, 40–43

D
Danckwerts, surface renewal theory of, 59
Density, 10
Dew point, 11
Differentiation, 81
Diffusion
  equimolar counter-, 56, 61
  in solids, 63–64
  radial, 63
  steady state, 55

Diffusivity
  binary, 54–55
  in multi-component mixture, 55
Dimensions, 1–4
Dulong's formula, 20

E
Energy
  heat, 18
  internal, 17
  kinetic, 17
  potential, 17
Energy balance, 10, 17, 29
Energy balance, and reactor design, 82
Enthalpy, 18–19
  of a solution, 43–44
Entropy, 30–31
Equations
  adiabatic process, 33
  Antoine, 12
  Beattie-Bridgeman, 14, 35
  Benedict-Webb-Ruben, 14
  Clapeyron, 39–40
  Clausius-Clapeyron, 12
  efficiency, 31
  energy absorption, 32
  energy balance, 30
  entropy, 31–33
  Frossling, 47
  Gibb's free energy, 32
  Helmholtz function, 32
  isobaric, 33
  isometric, 33
  isothermal, 33
  mass balance, 16–17, 30
  Michaelis-Menton, 80
  performance, 82
  polytropic compression, 33–34
  Redlich-Kwong, 14, 35
  of state, 12
  Stefan-Maxwell, 55
  two constant, vapor pressure, 12
  Van der Waals', 13–14, 34–35
  van't Hoff, 72
  virial, 14
  Watson's empirical, 12
  Wilke and Chang relation, 55
Equilibrium
  chemical, 71–72
  constant, 71–72
  conversion, 71, 73

F
Flow rate, 11
Flow reactors, 83–85
  continuous, stirred tank, 84
  mixed, in series, 84–85
Force, 1–2
Formation, heat of, 70
Freezing point, depression, 15
Fuels, 19–20
Fugacity, 35–38
  coefficient, 35
  in ideal solutions, 36–37
  and phase equilibria, 37–38

G
Gas
  ideal, 12–13, 33–34
  non-ideal, 13–14
  real, 13–14, 34–40
  saturated, 15
  thermodynamic relations for, 33–34
  unsaturated, 15–16
  and water vapor mixtures, 16
Gibbs' phase rule, 44

H
Half-times, 81
Heat
  of chemical reaction, 18, 70
  of combustion, standard, 19, 70
  energy, 18
  of formation, 18, 70
  of mixing, 43–44
  of reaction, standard, 18, 19
  of vaporization, 12
Heating value, 19
Henry's law, 37
Hess, Law of, 19
Higbie penetration model, 58–59
Homogeneous reactions
  catalyzed, 77
  reversible, 74–76, 79
Humidity, 15–16
Hydrate, 16

I
Investment
  fixed, 94
  return on, 94

K
k calculation, 81
Kinetics, Chemical, 69–85

L

Law
- Amagat's, 13, 33
- Boyle's, 12, 33
- Charles's, 12, 33
- Dalton's, 13, 33
- Fick's, 53–54, 63
- of Hess, 19
- mass action, 74
- Raoult's, 14
- of thermochemistry, 19

Lewis-Randall rule, 37

M

Mass action, law of, 74
Mass balance, 10, 16, 29
- and reactor design, 82

Mass fraction, 10
Mass percentage, 10
Mass transfer, 53–64
- coefficients, 56, 60–61
- film model, 58
- from gas to liquid film, 57
- interphase, 59–60
- under non-flow conditions, 55–56
- in packed beds, 61–63
- from spheres, 57–58
- turbulent, 58

Methods of Analysis, 81
Michaelis-Menton Equation, 80
Mol fraction, 10–11
Molality, 11
Molar volume, 13
Molarity, 11
Mole (mol), 2
Molecularity, and order of reaction, 74
Mollier diagram, 39
Multipliers, 2–4

N

Normality, 11

O

Optimization, degrees of freedom for, 93
Order-of-magnitude estimate of costs, 95–96
Order of reaction, 74

P

Payback time, 95
Performance equation, and reactor design, 82

Plug-flow reactors, 84
Pressure, critical, 13
Process
- irreversible, 28
- reversible, 28

Profit, 94
- venture, 95

Properties
- extensive, 28
- intensive, 28
- state, 28–29
- thermodynamic, 28–30

Psychrometry, 16

R

Rankine cycle, 43
Rate expressions, 78–79
Reactions
- autocatalytic, 78
- complex, 76
- consecutive, 77
- effect of temperature on, 82
- homogeneous, 73
- homogeneous, catalyzed, 77
- irreversible, 74, 78–79
- parallel, 76–77, 79

Reactors
- batch, 82–85
- continuous-flow stirred tank, 84
- design for homogeneous reactions, 82
- flow, 83–85
- mixed-flow, in series, 84–85
- plug flow, 84
- steady-state tubular, 84

Reduced conditions, 38–39
Reference curves, 81
Refrigeration, 41

S

Saturation, 15–16
- temperature, 11

Solubility, 16
Solutions [chemical], 14–15
Solvates, 16
Solvent
- extract, 16
- raffinate, 16

Space-time, as performance measure, 83
Space velocity, as performance measure, 83
Specific gravity, 10
Specific volume, 10

Standard state, 37
State, corresponding, 13, 39
Steady state condition, 10
Stefan-Maxwell equation, 55
Surface renewal theory of Danckwerts, 59
System
- closed, 28
- open, 28

T

Temperature
- critical, 13
- dry-bulb, 16
- wet-bulb, 16

Theorem of corresponding states, 39
Thermochemistry, law of, 19
Thermodynamics, 27–44
- of chemical reactions, 70–71

Thermodynamics, laws of, 27–40
- First, 29–30
- Second, 30–33
- Third, 40

Transfer
- energy, 29
- mass, 29

Turbine expansion cycle, and vapor compression, 42
Turbulent boundary layer theory, 59

U

U. O. P. characterization factor, 20
Units, 1–4
- derived, 1
- fundamental, 1

V

Vapor compression with free expansion, 42–43
Vapor pressure, relative, 14–15
Venture cost and profit, 95
Volume, critical, 13

W

Water vapor, and gas mixtures, 16
Wilke and Chang relation, 55
Work, 18